Zhen Chuan Kang

Electrons and Electron Microscopy

Also of interest

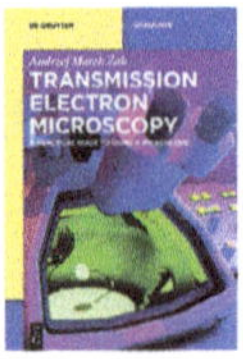

Transmission Electron Microscopy.
A Practical Guide to Using a Microscope
Żak, 2024
ISBN 978-3-11-131649-9, e-ISBN 978-3-11-131701-4

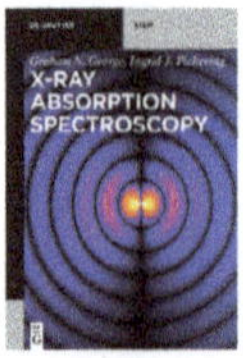

X-ray Absorption Spectroscopy
George, Pickering, 2025
ISBN 978-3-11-057037-3, e-ISBN 978-3-11-057044-1

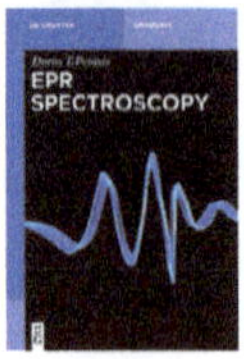

EPR Spectroscopy
Petasis, 2022
ISBN 978-3-11-041753-1, e-ISBN 978-3-11-042357-0

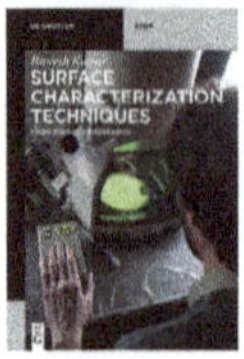

Surface Characterization Techniques.
From Theory to Research
Kumar, 2022
ISBN 978-3-11-065599-5, e-ISBN 978-3-11-065648-0

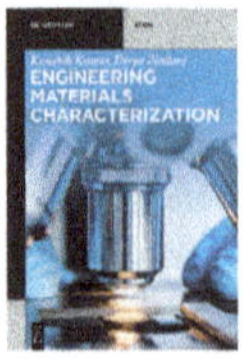

Engineering Materials Characterization
Kumar, Zindani, 2023
ISBN 978-3-11-099760-6, e-ISBN 978-3-11-099759-0

Zhen Chuan Kang

Electrons and Electron Microscopy

Quantum Electron Microscopy Introduction

DE GRUYTER

Author
Dr. Zhen Chuan Kang
Cedar Park, TX, 78613
United States of America

ISBN (PAPERBACK): 978-3-11-144913-5
e-ISBN (PDF) 978-3-11-144933-3
e-ISBN (EPUB) 978-3-11-144969-2

Library of Congress Control Number: 2024943783

Bibliographic information published by the Deutsche Nationalbibliothek
The Deutsche Nationalbibliothek lists this publication in the Deutsche Nationalbibliografie; detailed bibliographic data are available on the Internet at http://dnb.dnb.de.

Cover image: Mohammed Haneefa Nizamudeen/iStock/Getty Images Plus
Typesetting: Integra Software Services Pvt. Ltd.

www.degruyter.com
Questions about General Product Safety Regulation:
productsafety@degruyterbrill.com

This book is dedicated to the memory of
Professor John Maxwell Cowley and Professor Leroy Eyring

About the book

Recent advancements in Transmission Electron Microscopy are built upon the remarkable achievements of the transmission electron microscope, especially with the aberration-corrected objective lens. This lens represents the incoherent integration of particle electron optics and modern wave imaging technology, embodying the electron's particle-wave duality. However, questions arise: Why is the contrast of the images on micrographs interpreted as a mass-thickness difference (called mass-thickness contrast), energy distributions between the transmitted and scattered electron beams (called diffraction contrast), or planar electron wave interference (called phase contrast)? Given that a planar electron wave disseminates in all directions, how can it simultaneously exist as a charged particle at a defined location?

These contradictions and confusions lead us to fundamental inquiries: What is an electron? What is electron microscopy? Why can the principles of optics for optical microscopes be applied to bring success to electron microscopy? This book attempts to answer these questions by applying de Broglie's Hypothesis and Einstein's Theory of Relativity to explore the relationship between particles and electromagnetic waves, offering insights into the field of electron microscopy. It provides an introduction to what quantum electron microscopy can be, potentially advancing the field of microscopy beyond atomic spatial structural periodicity.

- Refreshes the concept of high-resolution imaging. (from refreshes the idea of the high-resolution images)
- Explains the wave-corpuscle duality of the electron. (from explains the duality of the wave-corpuscle of an electron)

https://doi.org/10.1515/9783111449333-202

Preface

Almost 40 years ago, when I first came to the United States to learn transmission electron microscopy with Professor John Maxwell Cowley at Arizona State University, I knew little about electron microscopy. The little knowledge I had about electron microscopy was taught by Professor Huang Lanyou, who was one of the students of Professor Möllenstedt. When I worked for Professor Leroy Eyring to solve the relationship between fluorite-related superstructures and oxygen vacancy configurations of non-stoichiometric oxygen-deficient rare earth higher oxides, I tried to use transmission electron microscopy to decipher the configuration rules of the oxygen vacancies in the rare earth higher oxides with fluorite-related structures without success. However, using neutron diffraction technique solved the problems. Professor Leroy Eyring and I went to the office of Professor John M. Cowley to discuss these problems. He gave us some advice and introduced us to the book *Atlas of Optical Transformations* written by G. Harburn, C.A. Taylor, and T.R. Welberry. Professor Leroy Eyring and I discussed questions such as: what is an electron? It is a particle or a wave? Why is a plane wave function usually used to express the electron in literature of the electron microscopy? These questions perplexed us in understanding the high-resolution electron micrographs collected from the transmission electron microscope.

The books such as *High-Resolution Electron Microscopy* written by J.C.H. Spence, *4D Electron Microscopy Imaging in Space and Time* written by Ahmed H. Zewail and John M. Thomas, and *Advanced Transmission Electron Microscopy* written by J.M. Zuo and J.C.H. Spence comprehensively review and expound on the advanced transmission electron microscopy. The questions I had still haunt my dreams. However, a recent paper by Professor A. Howie mentioned that "In his (Ruska's) Nobel prize acceptance speech, Ernst Ruska admitted that he became aware of the wave properties of the electron only as late as 1931 when he had already made his invention." The success of modern electron microscopy is based on wave imaging technology. The future of transmission electron microscopy will be shaped by "continued skirmishing on the wave-particle frontier" (Howie, 2019). The idea of Professor A. Howie encourages me to write this book, *Electron and Electron Microscopy: Introduction to Quantum Electron Microscopy*.

I wish this book to be a memorial for the 100th anniversaries of Professor Leroy Eyring and Professor John Maxwell Cowley, who significantly contributed to the advanced electron microscopy in the last century.

https://doi.org/10.1515/9783111449333-203

References

Howie, A. “Continued Skirmishing on the Wave-Particle frontier.” Ultramicroscopy, 203, 52–59, (2019).
Spence, J.C.H. High-Resolution Electron Microscopy. 3rd Ed, New York: Oxford University Press, (2003).
Zewail, A.H. and Thomas, J.M. 4D Electron Microscopy Imaging in Space and Time. Imperial College Press, (2010).
Zuo, J.M. and Spence, J.C.H. Advanced Transmission Electron Microscopy. Springer, (2017).

Contents

Chapter 1
Understanding the electron is essential physics of electron microscopy

1.1 Discovery of electrons and invention of the microscope

In 1897, J.J. Thomson discovered the first "elementary particle," electron, by measuring its charge-to-mass ratio (e/m), which was crucial for understanding the nature of the electron and atom. Thomson experimentally manipulated the trajectories of the moving electrons in electric and/or magnetic fields and using Newton's equations of classical mechanics calculated the balance of the acting forces on the electron. He discovered that the electron was a particle with mass ($m = 9.1 \times 10^{-31}$ kg) and charge (1.6×10^{-19} C).

In 1923, Prince Louis de Broglie postulated that ordinary matter can have wave-like properties, with the wavelength λ related to momentum p as $\lambda = h/p$ (h is Planck's constant, $h = 6.63 \times 10^{-34}$ Js). In 1927, the Davisson-Germer's experiment, in which an accelerated electron beam scattered from a nickel (Ni) crystal, found several sharp peaks in intensity related to a function of the accelerated voltage of the electrons. In 1928, C.J. Davisson published the paper titled, "Are electrons waves?"

In the same year, Hans Busch discovered that the magnetic field of a solenoid acting on an electron beam results in the same way as a glass lens acting on a light ray. In 1931, E.A.F. Ruska realized that using an electron beam, instead of a light ray, to build a microscope might yield better resolution than an optical microscope and successfully invented the first electron microscope. However, he did not know about de Broglie's matter wave idea.

Since Ruska invented the electron microscope, electron optics and electron microscopy have made tremendous advances. For instance, today the aberration-corrected scanning transmission electron microscope (STEM) may have better than 0.1 nm lattice image resolution and clearly shows the atoms' spatial periodicity in the projected structure of a crystal along with the electron beam direction. However, understanding the physical meaning of the observed images is still not fully convincing, and basic questions are still debated such as whether electrons are particles, waves, or both.

It is well known that an electron demonstrates the characteristics of a particle (de Broglie called it a "corpuscle") with mass, charge, and spin, but it also exhibits electron-positron annihilation creating γ-ray. When moving electrons pass through slits, an interference pattern occurs. These properties exhibit the wave nature of the electron, which is the same as an electromagnetic wave or light wave. Then what is an electron?

Ruska's electron microscope should be the best instrument to discover what an electron is because the interaction between matter and electrons and the electron

https://doi.org/10.1515/9783111449333-001

beam imaging process should provide enough evidence to unveil the electron's intrinsic nature.

In 1955, Gottfried Möllenstedt and Heinrich Düker used the electron biprism to obtain interference fringes with an electron microscope. In 1961, Claus Jönsson performed a five-slit electron interference experiment. In 1976, Merli, Missiroli and Pozz published a paper, "On the statistical aspect of electron interference phenomena," and a 16-mm movie showing sequential arrival of electrons, one at a time, on a television monitor. Using single electrons, in 1989, Akira Tonomura conducted a double-slit interference experiment (using biprisms as double slits) in a Hitachi electron microscope, which was called "the most beautiful experiment in physics" in the twentieth century, and demonstrated that the electrons have both particle and wave nature. In 2011, Sacha Kocsis et al. published the trajectory of a single photon in a two-slit interference experiment. In 2013, Roger Bach et al. published a controlled real double-silt (instead of biprisms) single-electron diffraction experiment. In 2015, using a transmission electron microscope, Fabrizio Carbone obtained a snapshot image of the photon. In 2016, Reuben S. Aspden et al. published, "Video recording true single-photon double-slit interference." All of these experiments demonstrate that electrons and photons have the characteristics of both particles and waves. Figure 1.1 shows the results of Tonomura's experiment, and Figure 1.2 shows the results of an experiment by Bach et al. (The figures are the edited selected copies.)

The electron detection rate was about 1 Hz, and the average distance between consecutive electrons was 2.3×10^6 m, which ensured that only one electron was in the 1-m-long system at any given time and eliminated electron-electron interactions. Figure 1.2 shows the edited diffraction patterns. Over time, the electron, as a particle, randomly hits the recording detector (Figure 1.2a–c). Only after considerable time had passed and enough electrons (more than 10^5 electrons) had accumulated, did the usually observed diffraction pattern develop as shown in Figure 1.2d. The final built-up diffraction pattern (Figure 1.2e) took about 2 h.

On the basis of these experiments, we have to believe that an electron has both particle and wave nature, but not as a particle or a wave. In the general literature about electron microscopes, electron optic analysis of electron guns or magnetic lens systems treats electrons as charged particles. However, for the description of image formation of electrons, electrons are often expressed as a plane wave. This implies that electrons have either a particle or a wave nature in different situations.

What is the difference between "particle and wave" and "particle or wave"? Is it important for electron microscopy?

1.1.1 Is an electron is a particle or a wave?

Based on the orthodox interpretation of quantum mechanics, in Bohr's words, electrons are associated basically with probability waves whose modulus of the amplitude gives

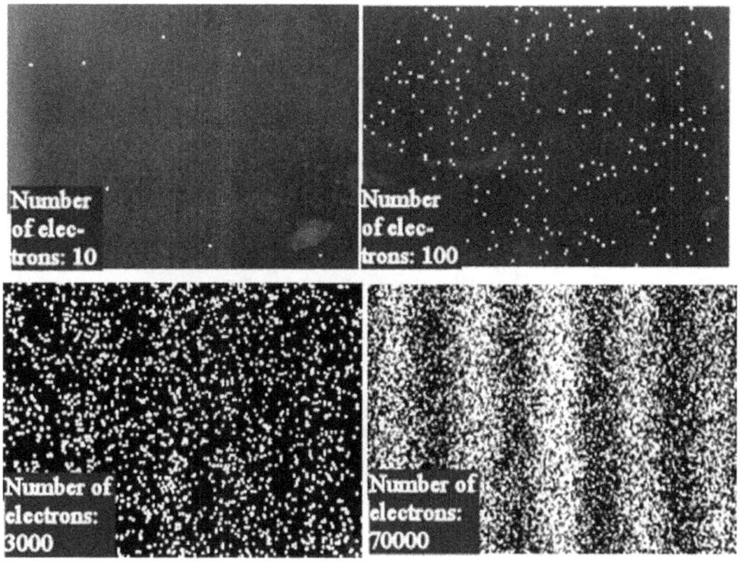

Figure 1.1: Single-electron double-slit (biprism instead of real slits) interference, "the most beautiful experiment in physics," shows the particle and wave nature of electrons.

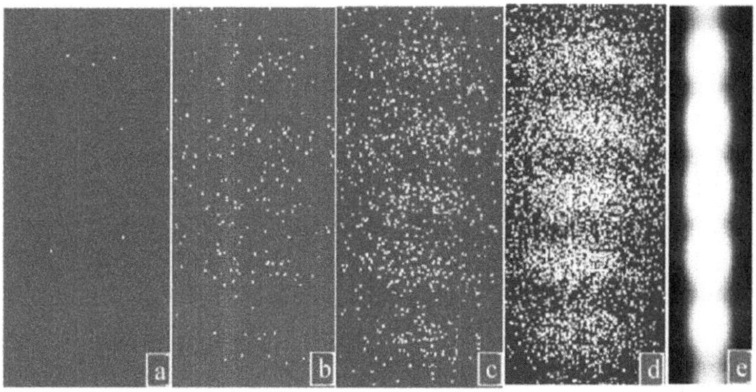

Figure 1.2: Single-electron double-slit (two 62 nm slits, with a center separation distance of 272 nm) experiment pattern. Figure 1.1(a)–(d) shows progressive snapshots and Figure 1.2(e) shows electron behavior, recorded after 2 h.

the probability density of finding an electron particle. The particle character of the electron appears when one measures the position of the electron. The electron cannot be both a wave and a particle at the same time; it must be either one or the other depending on the situation. For the double-slit experiment, the electron primarily presents as a wave with a plane wave function traveling through both the slits. When a position measurement is taken, the wave function suddenly collapses into a delta function at a random, particular position on the recording screen. The particle-like nature of the electron appears only when its wave-like nature disappears. Since the screen positions where collapse occurs follow the probability distribution dictated by the modulus of the wave function, a wave interference pattern appears on the detector screen.

1.1.2 Electron is both a particle and a wave

de Broglie and Bohm presented an explanation of quantum phenomena: when the concepts of particle and wave merge at the atomic scale (smaller than 10^{-13} cm), they assume a pilot-wave guiding the motion of the electron, but the electron is a particle. Therefore, an electron is a particle having a pilot wave that guides its motion by the Schrödinger equation.

The de Broglie and Bohm theories define that the amplitude modulus of a wave functions as the particle probability density of being at a certain position, regardless of the measurement process. In the double-slit experiment, the wave travels through both slits. At the same time, a well-defined trajectory is associated with the electron. Such a trajectory passes through one of the slits. The final position of the particles on the recording screen and the slit through which it passes are determined by the initial position of the particle, which the experimenter cannot control, and so there is an appearance of randomness in the pattern of detection. The wave function guides the particle in such a way that it avoids those regions, in which the interference is destructive and is attracted to constructive interference, giving rise to the interference pattern on the detecting screen as shown in Figure 1.3.

We may note that those single-electron double-slit experiments support the pilot-wave idea. As John Bell said, "The guiding wave, in the general case, propagates not in ordinary three-dimensional (3D) space but in a multidimensional configuration space that is the origin of the notorious "non-locality" of quantum mechanics. It is a merit of the de Broglie-Bohm version of quantum mechanics to bring this out so explicitly that it cannot be ignored." Hence, it seems that using the pilot-wave theory may be better to unveil the mysterious quantum phenomenon for solid and biomaterial by using a single electron, an electron bunch, or an electron beam in the electron microscope.

Therefore, understanding the characteristic features of particles and waves may be the first thing to do.

1.2 Shape and structure of an electron

Frank Wilczek indicated what an electron is: an electron is a particle and a wave; it is ideally simple and unimaginably complex; it is precisely understood and utterly mysterious; it is rigid, and subject to creative disassembly. No single answer does justice to its reality.

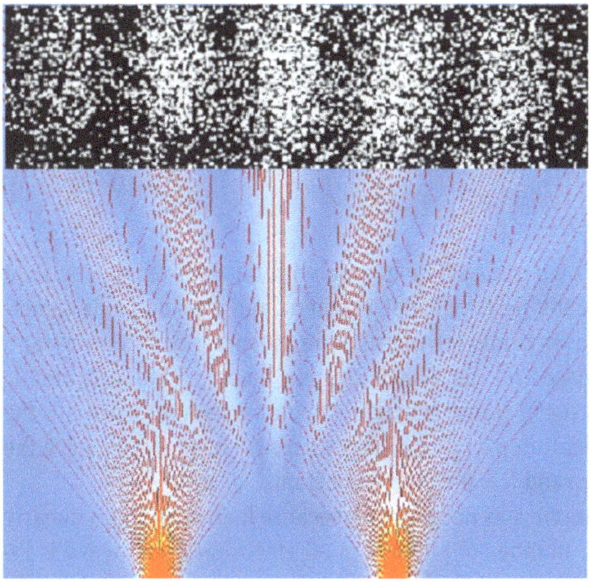

Figure 1.3: de Broglie-Bohm theory explains the interference of double slits, using the electron trajectories guided by the wave function in the configuration space.

Dirac and Einstein recognized in the 1920s that all matter is no more than localized electromagnetic energy. If we accept this idea, then electron would be a localized electromagnetic energy in a corpuscle, as de Broglie said. Recent advances in understanding the nature of an electron seem to prove the idea of Dirac and Einstein. A matter wave, such as an electron, is "confined single harmonic electromagnetic waves," which is the de Broglie wave of matter. J.G. Williamson's model of the electron as a dynamic photon with toroidal topology based on the super-relativity theory may be useful to understand the intrinsic characteristics of electron and could be consistent with de Broglie-Bohman quantum mechanics. This model of the electron may serve as the basis for understanding what has been seen with electron microscopy and what might be developed with it.

1.2.1 The electron model of a dynamic photon with toroidal topology

If the idea of Dirac and Einstein is accepted for forming a photon or an electron, the electromagnetic energy must be confined in the curved metric area with an internal frame with its own space-time metric. J.G. Williamson et al. used extended Maxwell's equations and Newton's laws to prove that the electromagnetic field can be confined to a confined space with the balance of electromagnetic and gravitational interaction that emanates from the absolute relativity of the rotating motions of an electromagnetic vorticity of a photon in space and time. Using a similar method, J.G. Williamson proposed that the model of an electron has a torus morphology that makes the particle feature of the electron nestling a space-time geometry frame or its own Minkowski space.

We use this model to understand the particle and wave nature of an electron.

1.2.1.1 What is a photon?

A photon is a particle (as a tiny corpuscle) or quanta with an electromagnetic field energy that contains the mutually perpendicular magnetic and electric fields, generally, changing in the harmonic ways and having a certain polarization of electric and magnetic fields in common. Based on the Planck-Einstein theory, a monochromatic electromagnetic wave will consist of N monoenergetic photons, each of which has the following energy ε_p and momentum p_p: $\mathcal{E}_p = h\omega = \hbar\nu$, where $\hbar = h/2\pi$, ν and ω are the linear and circular frequencies, and h is the Planck constant.

The photon, as a tiny corpuscle, has no rest mass because it always moves linearly with the speed of light, c. The number of photons in an electromagnetic wave is related to their total energy: $E = N\mathcal{E} = N\,\hbar\nu$. The angular moment or spin of one photon is equal to $J_p = \hbar$. All photons have identical energy and polarization. Because a photon is characterized as having only one frequency, any photon is regarded as a monochromatic or monoenergetic particle. The energy of a photon is given as $\mathcal{E}_0 = \hbar\,\nu = \hbar\,c/\lambda$, which implies that the energy of a photon may be concentrated within the space of a wavelength scale, the smallest part of an electromagnetic wave, called a "photon" or an "energy packet." Monochromatic traveling electromagnetic waves consist of the integer numbers of wavelengths as shown in Figure 1.4.

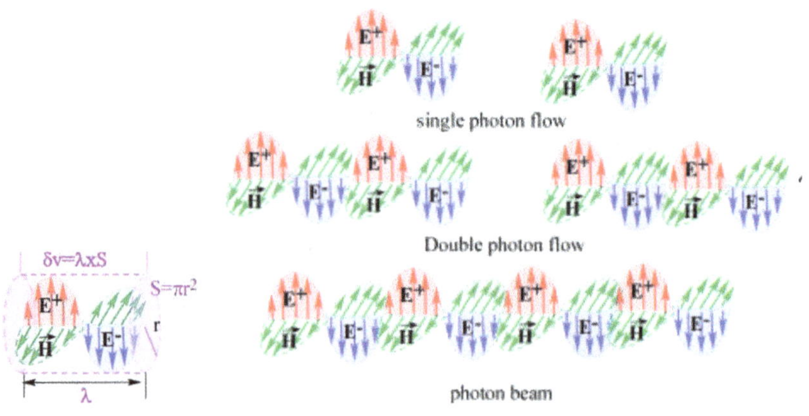

Figure 1.4: Quanta photon and the electromagnetic wave.

Each photon should have the identical electromagnetic energy, expressed as

$$U_0 = (1/8\pi)\int (E^2{}_0 + B^2{}_0)\,dV$$

This electromagnetic energy does not clearly indicate the oscillation of the electric and magnetic fields. However, the Planck constant gives the photon energy as

$$\varepsilon_p = \hbar\nu$$

If the ν is one circle per second, then the Planck constant, ħ, would only be the electromagnetic energy in this circle or in the lowest natural limit of the frequency. It is the quantum limit or state of frequency. The dimensions of h are gm-cm^2-s^{-1} or gm-cm^2-t^{-1}. Frequency is 1/T, where T is the oscillation period; gm-cm^2-sec^{-1} is the energy density, which causes an action or resulting in an effect. Therefore, the Planck constant h may be seen as the unit of action or effect, and the unit actually is a quantum, the minimum and smallest value of the electromagnetic energy.

If we use the de Broglie relation $p = hk = h/\lambda$ and the linear momentum in Newtonian mechanics $p = mv$, then $h = m\lambda v = mA/t$, where A is the area. Therefore, h is some mass moving with the light speed through some area during an elapsed time. The sweep area is a constant: h/m. This fixed area is swept out by the de Broglie wavelength of the particle (photon or electron) at either any steady or a varying speed in conjunction with any singular or compounding paths in one, two, or three dimensions, which may be a straight line, zigzag, circle, elliptical, spiral, helix, or a whirled, winding, or random (chaotic) configuration within a definite range of time.

If this analysis is acceptable, then a photon may have the configuration of a flat torus, because if $h = m\lambda^2\nu$ and λ may be seen as the radii of the torus R, then $h = mR^2\nu$; mR^2 is the moment of inertia of the ring; and $mR^2\nu$ is the angular momentum of the rotating ring. Therefore, the configuration of a photon is the electromagnetic field, $(E_0 + B_0)$ rotating (right or left hand) with the speed of light around the Poynting vector $P = (cE_0 \times B_0)$, which is the flux direction of electromagnetic energy density or the electromagnetic momentum density $G = 1/c\ (E_0 \times B_0)$ as shown in Figure 1.5.

Based on the theory of special relativity, no particle can rotate with a tangential linear velocity greater than the velocity of light, so the angular momentum of a photon is

$$L = mR^2\nu = mRc = s$$

which is the spin of a photon; $m = h/cR = h/c\lambda$ is the equivalent rest mass of a photon. If we postulate a negative mass as a black hole, the photon may have a black hole as $-m = h/c\lambda$. The radius of the black hole may be the same as the space extension of a photon, called the "Schwarzschild radius": $r_b = 2\ Gm/c^2$ or $r_b/m = 2\ G/c^2$ where G is the gravitational constant. The Schwarzschild radius is the minimum radius that a photon can have. The maximum energy possibility for a photon will be

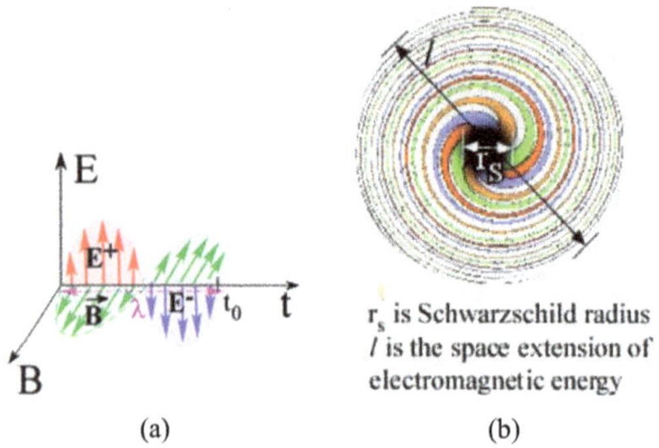

Figure 1.5: The configuration of a photon: (a) the electromagnetic field, with E and B, of a photon as a function of time; and (b) the electromagnetic field *e*nergy, $\varepsilon = \hbar\nu$, of a photon confined in the limited space by the wavelength.

$$mc^2 = \sqrt{\frac{hc}{2G}}c^2 = 8.61 \times 10^{22}\,\mathrm{MeV}$$

This is called "Planck energy," the highest energy of a photon. Until now, the highest observed photon energy is on the order of 10^{13} eV. We may indicate that $r = (2Gh/c^3)^{1/2}$ usually called "Planck length," which is the minimum length allowed for a photon. At the Planck length, the photon attains maximum energy and is transformed into a black hole. The Planck length is on the order of 10^{-35} m.

It is worth noting that the energy and mass of matter are two phases of the same physical entity. A recent theory proposes that the mass of a photon at rest is equal to m_0 (photon) = $h\nu/2c^2$, which is about 2.22×10^{-36} kg for a typical light frequency in the range of $4–8 \times 10^{14}$ Hz. As is well known, the mass of an electron at rest is 9.109×10^{-31} kg. If the frequency of a photon is 10^{20} Hz, its mass at rest may be similar to the mass of an electron.

1.2.1.2 What is an electron?

Based on Einstein's theory of relativity, the energy of a material particle is the product of its mass and the squared speed of light, expressed is $\mathcal{E} = mc^2$. In free space and in the observer's laboratory frame, the relationship between kinetic energy, $\mathcal{E}_T$, and momentum, p, is $\mathcal{E}_T = p^2/2m$, or $p = \sqrt{2m\mathcal{E}_T}$. This implies that the square-root mass-energy in any space-time frame may equal a physical parameter such as momentum p or wave vector k because p = h k. The square-root mass-energy density may be similar to the probability amplitude of the wave function in nonrelativistic quantum mechanics, expressed as Ψ, which when multiplied by an appropriate constant, squared and integrated over a relevant volume would give the energy of the electron in this volume. J.G. Williamson's group used the square-root mass-energy density with the electric field density, $1/2ED = 1/2\varepsilon_0E^2$, and magnetic field energy density, $1/2BH = 1/2\varepsilon_0c^2B^2$ (ε_0 is the electric permittivity and c is the speed of light) to solve the Maxwell equations concerning Newton's law. The results unveil the quantization of the electromagnetic field and the nature of the photon and the electron. J.G. Williamson's results are consistent with previously proposed models of the electron and recent theory, but they elucidate more fundamental physics comprehensively.

As is well known, mass represents the measurement of inertia or momentum that is changed by the action of a force. The rest-mass field allows any incoming moving field to merge into a re-circulating vortex-like motion to localize the moving photon in a minimum configuration space (Minkowski space). To create the minimum energy configuration, the momentum of the square-root rest mass energy density should rectify the momentum (or Poynting vector) of the electromagnetic field energy flux of the photon moving at the speed of light into a curling configuration. Therefore, the electromagnetic field energy flux of the photon is thus confined by the mass, and the mass is also confined by the electromagnetic field, which implies a balance of the field and gravity in space and time, instantaneously changing momentum along a curling motion related to the mass action. This may be viewed as a rapidly rotating running sheepdog that, at light speed, corrals the rest-mass component that is its attractor. The distribution of momentum of the square-root energy density dominates the shape and extension of the path of the moving photon at light speed and the photon has a spin of h (Planck constant), which indicates the rotating electric and magnetic fields as shown in Figures 1.5 and 1.6. Therefore, the moving photon has two types of motion: one is the spin in the photon's own frame as vortex-like, and the second is the circulating helix motion in the observer frame. The photon moving at the speed of light circulates around the two frames, giving rise to the topologic morphology of a torus, as shown in Figure 1.6. In momentum space, the momentum, p, distribution, raised by the square-root mass-energy density, is a distorted sphere, which relates to the torus parameter of r_2/r_1.

Therefore, to form an electron as a spatially confined particle or corpuscle, the electromagnetic energy flux must be localized in the curved metric area. This implies that the strong force must act circumferentially in a two-dimensional curved domain, rotating close to the velocity of light around two orthogonal directions. Both directions are orthogonal in the torus, which is its own Minkowski space, and the confined electromagnetic energy flux as a corpuscle may also have a trajectory in the laboratory observer space. The r_2 relates to the observer laboratory space, and r_1 is related to the spin of the vortex photon that is its own internal space frame.

Therefore, a resting electron is a corpuscle containing dynamic electromagnetic field energy flux in its limited, confined, internal frame space (or its own Minkowski space). The circulating electromagnetic energy flux in the corpuscle's internal frame experiences energy and momentum conservation, equalizing the momentum of the electromagnetic energy flux and the momentum of dynamic mechanical motion of the photon but the direction of these momenta is reversed while the corpuscle is stationary.

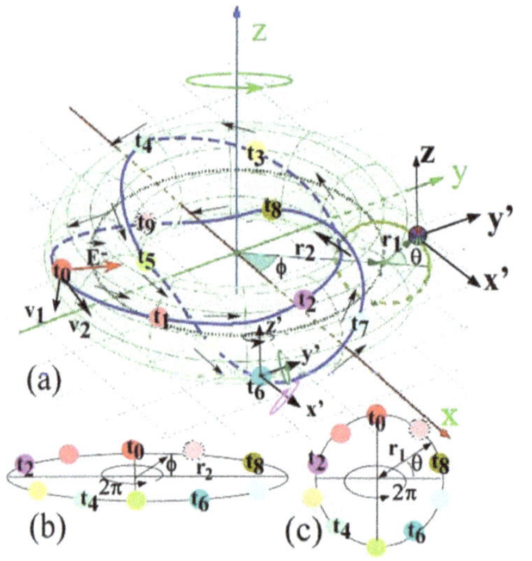

Figure 1.6: The photon path on the surface of a torus. The photon's own frame is z-x′-y′, and the laboratory frame is xyz. The photon rotating around in its own frame is z-x′-y′, and its own frame circulating around the z-axis in the laboratory frame is x-y-z.

Therefore, the particle behavior of the electron originates from the confined electromagnetic energy flux of the circulating photon, and the electron is a physical entity consisting of confined electromagnetic vorticity energy.

1.2–1.3 Spin and charge of an electron

(a) The spin of an electron

As mentioned before, a photon is a transverse electromagnetic field with a wavelength in the propagation direction. The electric field vector of a photon is the origin of the electric field of a charged electron.

The interacting process between the mass-energy and the electromagnetic field energy flux of the photon creates the confined, stable, topological configuration of the electron, occupying a closed-curve spatial domain in the observer laboratory frame. In this domain, the electromagnetic field of the photon can move along the closed curvilinear trajectory as shown in Figure 1.6. The transverse electromagnetic field is characterized by electromagnetic momentum, $P = (cE_0 \times B_0)$, Poynting vector and the circular electromagnetic momentum flux carry the energy flux, which will raise a characteristic electric current. When the electric field of the photon circulates around a circle with its radius r, then, based on the Maxwell equation, the electric field changing with time is

$$\partial E/\partial t = -(\partial E/\partial t)n - E(\partial n/\partial t)$$

as shown in Figure 1.7.

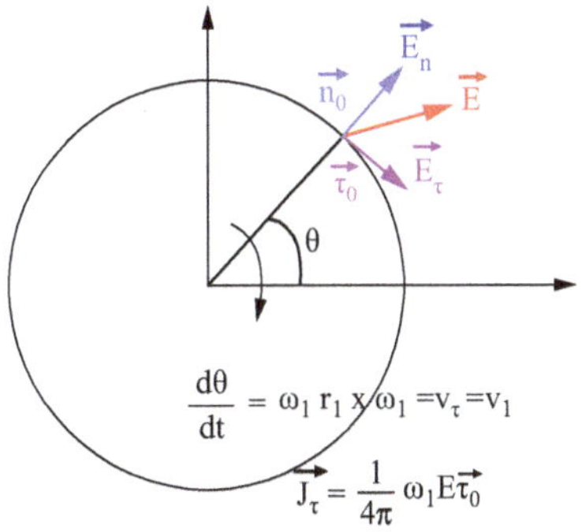

Figure 1.7: The electric field component of a photon circulating in its own frame induces a displacement current J_τ.

On the right-hand side of the equation, the first term indicates the electric field change with time along the direction of the radius. The second term shows that as the electric field of the photon twirls, a supplementary displacement current appears. It is not difficult to show that it has a tangential direction to the circle:

$$\frac{\partial \vec{n}}{\partial t} = -\upsilon_p K\vec{\tau}$$

where τ_0 is the tangential unit vector, $\upsilon_p = c$ is the velocity of the photon, $K = 1/r_1$ is the curvature of its trajectory, and r_1 is the curvature radius. Thus, the displacement current of the circulating photon can be written in the following form:

$$\vec{j}_{dis} = -\frac{1}{4\pi}\frac{\partial E}{\partial t}\vec{n} + \frac{1}{4\pi}\omega_K E\cdot\vec{\tau}$$

where $\omega_K = \upsilon_p/r_1 = cK$ is the curvature's angular velocity. If, in the tangent components of the current, the angular frequency is the ω_K coming from rest mass $\omega = m_0c^2/h$ and r_1 is the radius of the torus, the displacement current would be $J = (\omega_k/4\pi)E\bullet\tau$, in which τ is the unit vector of the tangential velocity. This displacement current of the electric field of the circulating photon would cause the magnetic momentum of the twisted circulating photon (Maxwell's law). The electric field changing in a radiating direction does not induce any dynamic current because the average of the instantaneous current in one circulation should be zero.

The rotational energy is $\mathcal{E} = L\omega$, where L is the angular momentum and ω is the angular frequency. For a photon $L = h$ and the total energy of a photon is $\mathcal{E}_{ph} = \hbar\nu = h\omega$. The energy of a photon is entirely electromagnetic energy and this energy resides in its spin. The confined circulating photon has to travel around twice to complete its total path length $\lambda = 2\pi c/\omega$. The internal circulating frequency ω_1 in the photon's own internal frame is twice the photon's rotating frequency ω_2 in the observer's laboratory frame. The internal circulating energy is equal to the confined photon energy, $\varepsilon = \hbar\nu = h\omega_1$, and then the intrinsic angular momentum of the corpuscle or torus would be $L = h\omega_2/\omega_1 = h/2$, the spin of the electron. The confined circulating vorticity photon is called a "zitterbewegung particle" and is related to the electron's jittery motion (Zitterbewegung or trembling motion).

(b) The charge of an electron
As mentioned before, a photon is a transverse electromagnetic field perpendicular to the electromagnetic field energy flux (or Poynting vector). The electric field vector of a photon is the origin of the electric field of a charged electron. The interacting process between the mass-energy and the electromagnetic field of the photon traveling a periodic path creates the confined, stable topological configuration of the electron occupying a closed-curve domain of space as a corpuscle electron, in which the electromagnetic field energy flux of the photon can move along the closed curvilinear trajectory in its own space-time frame, as shown in Figure 1.8. The transverse electromagnetic field transforms into the circular electromagnetic field, which results in the characteristic electric field of an electron in the laboratory observer frame. Because the circulating frequency is so much higher (10^{20} Hz), the observer sees it as stable static electricity, as shown in Figure 1.8.

In a fixed XYZ coordinate system in the laboratory observer frame, the photon is circulating with radius r_1 and simultaneously rotating around the Z-axis. The electron has a radiated electric field, which is comes from the photon's electric field. To ensure the electric field vector is always in a radiated direction, the photon must twist itself 180° in its own frame while circulating. This makes the angular frequency of the photon around the direction, which is the tangent velocity of the rotation around its own axis in its own frame, two times higher than the angular frequency of rotation around the Z-axis, as shown in Figure 1.8.

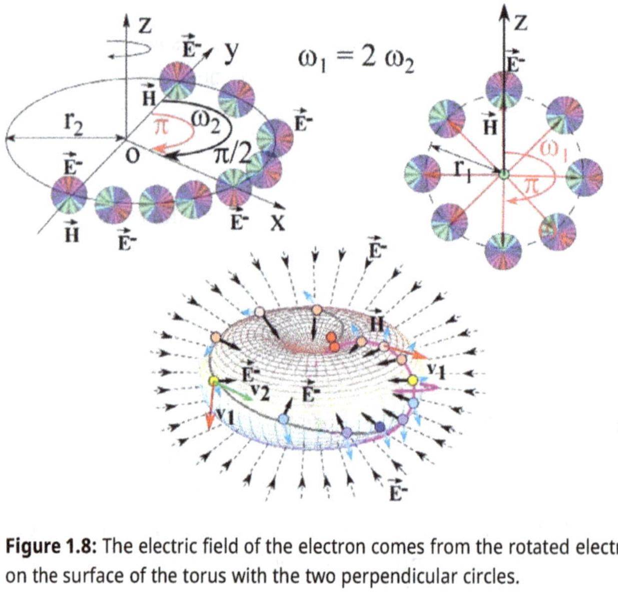

Figure 1.8: The electric field of the electron comes from the rotated electric field of the circulating photon on the surface of the torus with the two perpendicular circles.

The rest mass gives rise to the frequency and the radius of the photon, and so the angular frequency $\nu_1 = m_0c^2/h = 2\pi\omega_1$ and $r_2 = \lambda_{comp} = h/2\pi m_0 c$, which is the Compton length.

The field vectors of the photon have to rotate through 720° before coming back to their starting position with the same orientation. The circulating helix trajectory of the photon on the toroidal surface gives the periodicity of the path of the photon, but the radius of the torus is $r_2 = \lambda_{comp}/4\pi$. The origin of the photon's frame travels along the circumference of the torus at velocity v_2, but the photon is twisting and rotating in its own frame. The tangential velocity of the photon is v_1 and the internal circulating dynamic motions induce the spin of an electron and hold the energy of the rest mass. The rotation around the Z-axis is the orbital movement of a circulating photon containing the dynamic energy. From the laboratory frame view, the corpuscle occupies a spatial domain as a whole torus volume, which is a tiny space occupied by the particle with mass and spin and having stable static electricity.

The previous discussion depicts a distinct picture of what an electron is. The experiment shows the energy difference between the spin and the orbital movements in the electron's corpuscle frame, which proves the structure of the electron as shown in Figure 1.9.

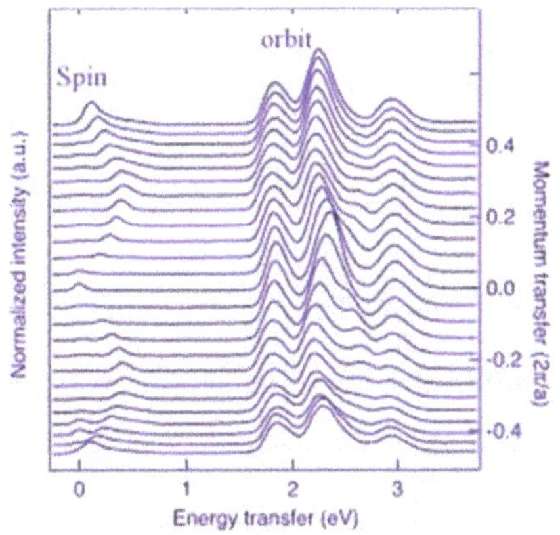

Figure 1.9: The energy difference between the spin and orbital movements of an electron in its own frame.

1.3 Description of a photon's dynamic motion at the spatiotemporal scope of a corpuscle electron

1.3.1 Electromagnetic field energy density

The electron is an elementary object that appears as a point particle with no structure in scattered experiments in the laboratory frame. It seems to be a very fundamental single entity. However, as we discussed previously, it may be a spatiotemporal dynamic system that implies (as in de Broglie's view) the electron holds its own source. The model of an electron proposed by J.G. Williamson, as we discussed in the previous section, may help us understand the nature of de Broglie's matter wave. A photon circulates at the speed of light around a main axis and simultaneously twists and rotates around its own axis, which is perpendicular to the main axis. This means that there are two frames. The first frame is referred to as a "static frame" or "main frame" related to the laboratory observer frame; the second frame is called the "internal frame," which rotates around the axis of the first one. In the second frame, the photon is traveling with twisting and circulating as it travels. The dynamic circulating process of the photon originates from rectifying the Poynting vector to produce a twisted, curved path of the momentum of the electromagnetic energy flux. The distribution of the square-root mass-energy density and the electromagnetic energy density dictates the topologic morphology of the corpuscle or the spatial distribution of the electric field of the electron. The toroidal radius of the rotating photon is r_2, and the radius of the twisting circulation of the photon at the speed of light in its own frame is r_1. The ratio of r_1/r_2 may be 1 or less than 1, $\zeta = r_1/r_2 \le 1$. When the electron is in a static state with the frequency of circulation of the photon $\omega_1 = \omega_0$, which is $\omega_0 = m_0c^2/h(=7.81\text{x}10^{20}\text{s}^{-1})$ and $\zeta = 1$ or $r_1 = r_2 = h/m_0c = R_c(=3.86\times10^{-13}\text{m})$ that is the Compton wavelength. The whole spatial domain of the torus contains the energy, $m_0c^2 = h\omega_0$, which is equal to the energy of the vortex photon. The electric charge distribution in this space is almost spherically radiated, and the electric field density may distribute as a Gaussian function. The momentum and energy of the photon play vital role in the construction of the weird torus topology of the electromagnetic energy of a photon.

If $\varepsilon_{Ein} = (m^2 + p^2)^{1/2}$ and $k = m/\varepsilon_{Ein}$, then the instantaneous energy of the dynamic motion on the torus can be written as

$$\varepsilon_t = (r_2 - r_1/r_2)\varepsilon(\pi/2(1 - r_1{}^2/r_2{}^2)^{1/2})/(1 - v_t{}^2/c^2)^{1/2} m_t{}^0 c^2 \quad 0 \le r \le r_2; m_t{}^0 = 4\pi r_2{}^2\chi$$

where $m_t{}^0$ is the stationary mass of the moving torus and v_t is its instantaneous velocity ($r_1/r_2 = \zeta$). The energy of the resting state of the torus is $\mathcal{E}_0$, and χ is the elapsed mass density at the curled trajectory of a moving vortex photon on the surface of the torus. The stationary mass of the electron should be $m_t{}^0 = 4\pi r_2{}^2\chi$, which implies that the photon's twist, circling radius is the same as the rotation radius of the internal frame of the photon. That is, $r_2 = r_1$ or $\zeta = 1$. These formulas indicate that the electron's energy, as a topological torus formed by a dynamic photon moving with the curling helix trajectory, is related to the motion process of the photon. The radius ratio between the photon's rotation and its circulation in the internal frame dominates the energy distribution in its stationary space. In other words, the electric field distribution of an electron may change widely in the spatial scale depending on the radius ratio of the photon motion. The electron, as a torus of the dynamic photon, is able to gain energy by changing the velocity of the whole torus, thereby gaining the kinetic energy of the whole torus, and simultaneously changing its innermost structure as defined by the parameter r_1 and r_2. The structural changes of the torus may be understood in terms of the relativistic Lorentz transformation between different inertial frames. Indeed, when stationary, the torus has a maximum surface ($r_2 = 0$ is nearly a sphere) and the hole in the torus has not appeared. When the torus moves, its transverse size begins to contract, and a hole starts to form in the torus; its surface gets smaller and smaller. On the other hand, when the torus starts to move, its kinetic energy increases as $\mathcal{E}_{kin} = 1/2\ m_t^0 v_t^2$ where $m_t^0 = \gamma m^0$ (γ is the Lorentz factor $\gamma = \left(1 - v^2{}_t/c^2\right)^{-1/2}$). Thus, we observe an important fact: the energy and electric field of an electron simultaneously depend on its velocity and its spatial dimension, which could be contractible or expandable, originating from the internal dynamic motion of the circulating photon.

The electromagnetic energy change accompanies the radiation of the electromagnetic wave. As is well known, the photon is the transfer unit of single quanta of electromagnetic energy and angular momentum between an electromagnetic emitter and an absorber. It is neither a particle nor a wave. It is merely the exchange process of electromagnetic energy flow between a transmitter and a receiver by an intermediate electromagnetic vorticity field.

If these points that we made in previous sections can be accepted, the next questions are how to understand the motion process of an electron in the laboratory frame, and what is the relationship between the electron and its internal dynamic process of the vorticity photon. In other words, how to understand the de Broglie waves.

1.3.2 How to describe the dynamic photon on surface of the torus?

As we mentioned earlier, in the laboratory frame, a moving electron may be viewed as an energetic corpuscle, but in the electron's own frame or in the corpuscle space, a photon with monochromatic energy, $\mathcal{E}_0 = m_0 c^2$ or $h\omega_0$, is sustainably circulating around two perpendicular axes as shown in Figure 1.10. The tracks of the moving photon at every moment on the surface of the torus form a closed periodic path or periodic orbit. The energy and momentum are conserved in the dynamic process. The tracks' orbit in position space is closed, and in the position-velocity (or position-momentum) space, it is also closed. The track of the circulating photon may be seen as a point of concentrated electromagnetic field energy:

$$\varepsilon = (1/2)\ E^2 + (1/2)\ B^2$$

The spatial locus of the tracks of the circulating photon changes with time, and the direction of the electric field always holds the radiated direction.

The inertia momentum $p\ (q, \mathcal{E}) = (2m_0\mathcal{E})^{1/2}$ rectifies the electromagnetic field momentum direction (Poynting vector) on a circle with radius r of the photon (as shown in Figure 1.10) to be radiated. The frequency of this circulation is $m_0c^2/\hbar \approx 7.69 \times 10^{22}$ Hz. While the photon is traveling a periodic path on the surface of the torus, the mass and charge may be described as $\rho_m = m_0/4\pi^2R_r$ and $\rho_e = q_e/4\pi^2R_r$. The corpuscle electron is a closed dynamic system. The electromagnetic field energy density (u) and momentum (p_{em}) flowing along the path of tracks of the moving photon at a light speed c give rise to the electromagnetic energy and momentum flux ($\delta u/\delta t$). Because this corpuscle is a closed system, the electromagnetic and mechanical momentum should be balanced. As shown in Figures 1.10 and 1.11, the photon circulates with two perpendicular circles. If we choose the center of a torus as the origin of the coordinate system of the electron, as shown in Figure 1.12, then any track of the photon can be represented as $x = (R + r\sin\theta)\cos\varphi$, $y = (R + r\sin\theta)\sin\varphi$, $z = r\cos\theta$. It is more convenient to use the cylindrical coordinate system. Then

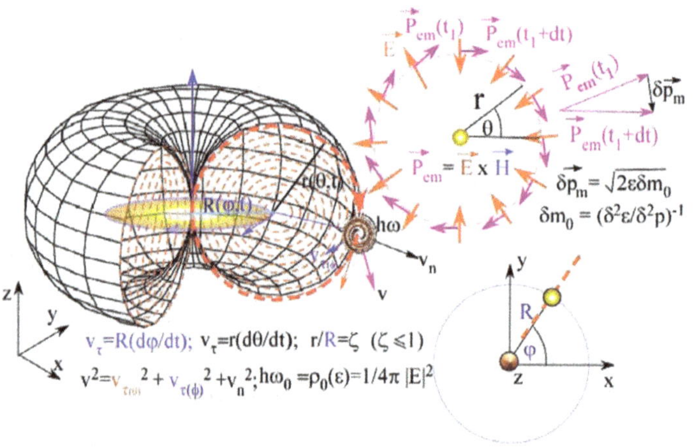

Figure 1.10: In the electron frame, the dynamic photon simultaneously circulates around two perpendicular circles. The tracks of the moving photon form a periodic curl path on the surface of a torus. The curl force curving the electromagnetic energy density flux is the momentum of mass-energy square-root flux. At each photon track, there are two tangential velocities perpendicular to each other.

$$r = R_0 + r_0\cos\theta,\ z = r_0\sin\theta,\ \upsilon_r = -r_0(d\theta/dt)\sin\theta$$

$$\upsilon_\varphi = r(d\varphi/dt);\ \upsilon_z = -r_0(d\theta/dt)\cos\theta$$

as shown in Figure 1.12.

The kinetic energy of the track with effective mass, ε/c^2, or the Hamiltonian is

$$H = \varepsilon = p_r{}^2/2m + p_\theta{}^2/2mr^2 + p_\varphi{}^2/2m(R_0 + r_0\sin\theta)^2$$

This Hamiltonian is separable, so the solution of the trajectory of the dynamically moving photon can be written as a product of three functions in the three coordinates, r,θ,φ:

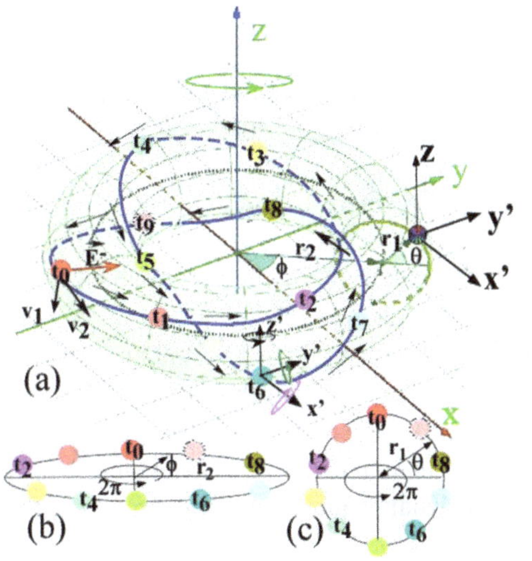

Figure 1.11: The tracks of a traveling photon along a periodic path on the surface of a torus at one periodic time $t_0 \approx 1.3 \times 10^{-21}$ s in the electron's own frame.

$$\Psi = \Psi_R \bullet \Psi_\theta \bullet \Psi_\varphi$$

Because the tracks of the circulating photon are always on the surface of the torus, the function Ψ_R has to satisfy the Nabla operator on Ψ_R to be zero:

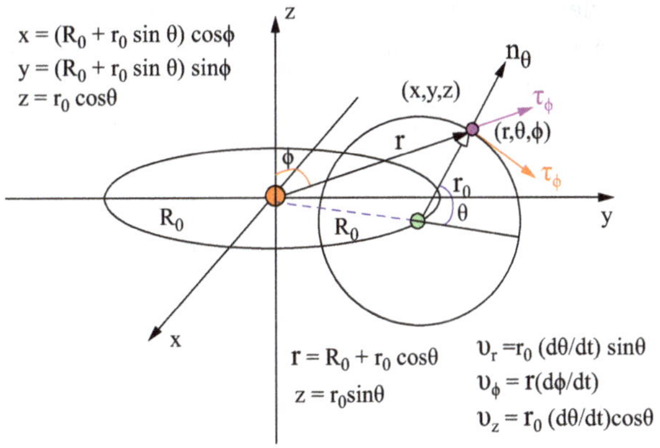

Figure 1.12: The cylindrical coordination and the track coordination in orthogonal coordination and cylindrical coordination.

$$\delta^2/dr^2(\Psi_R) + (1/r)(\delta/\delta r)(\Psi_R) = 0$$

so $\Psi_R = \kappa_0 \bullet \delta(r\text{-}\rho)$ would be a delta function. $\rho = 0$ is the origin of the torus, and the delta function can also be expressed as

$$\delta = 1/\sqrt{2\pi\alpha}\int \exp(-r^2/2\alpha)dr, \quad \text{where } 1/\alpha = k$$

$\Psi_R = \kappa_0 \bullet \delta(r\text{-}\rho)$ describes the shape or spatial extension for the corpuscle electron. Usually, it is called as the "spatial shape function."

The function Ψ_θ can also be obtained with $\Psi_\theta = A_0 e^{ikz}$ with $z = r\sin\theta$ and k would be $k = (2\mu\varepsilon/h^2)^{1/2} - (\mu^2/r^2)$ that is related to the energy $\mathcal{E}$ and the radius of vorticity photon. Then the function Ψ_φ is $\Psi_\varphi = B_0\exp\,(i\mu\varphi) = B_0\exp\,(i\mu\omega t)$, where μ is a chosen constant and it is related to the phase of the circulation of the vorticity photon.

Hence, the whole function describing the dynamic movement of the tracks of the circulating photon in the laboratory frame is

$$\begin{aligned}\Psi(r,\theta,\varphi) &= \Psi_R \bullet \Psi_\theta \bullet \Psi_\varphi = \kappa_0 \bullet \delta(r-\rho)A_0\exp(ikz))B_0\exp(i\mu\omega t)\\ &= \Psi_0[\delta(r-\rho)\exp(ikz)\exp(i\mu\omega t)]\end{aligned}$$

where $\Psi_0 = \kappa_0 A_0 B_0 = (4\pi^2 R_0 r_0)^{-1/2}$.

This is a functional manifestation of a corpuscle electron consisting of a twisted circulating photon with an electromagnetic energy density of $\mathcal{E}_0 = m_o c^2 = h\omega$. It shows that (a) an electron has its own shape as a small universe in the laboratory frame, appearing as a delta function. (b) Inside the electron's universe, there is a running sheepdog, which is a photon with electromagnetic energy $\mathcal{E}_0 = m_0 c^2$, which rotates at light speed c and corrals the rest-mass component that is its attractor. Therefore, the tracks of the twisted circulating photon create the effective mass of the electron and a periodic trajectory on the surface of a torus. (c) These periodic trajectories also create the radiated electric field or the electric charge and the spin of the electron. The function $\Psi\,(r,\theta,\varphi)$, usually called the "wave function," clearly demonstrates these intrinsic kinetic and dynamic nature of vorticity electromagnetic photon, which is generally called as a "corpuscle electron."

The electron is an energetic particle with $\mathcal{E}_0 = m_0 c^2$ and, simultaneously, is also an electromagnetic wave expressed as periodic electromagnetic field motion in the spatiotemporal scope. However, the Planck constant h is as indecipherable concealing an electromagnetic vorticity field.

Therefore, an electron as a wave-corpuscle is composed of confined dynamic electromagnetic wave energy. The wave function, $\Psi\,(r,\theta,\varphi) = \Psi_R \bullet \Psi_\theta \bullet \Psi_\varphi$, is the function describing the dynamic physical process of the corpuscle electron system. The wave function Ψ of an electron should not be understood as a plane wave or a probability wave.

1.3.3 Relationship between a moving corpuscle electron with velocity v_t and the internal dynamic process of a circulating vorticity photon in the corpuscle electron

The function Ψ (r, θ, φ), the so-called wave function, divulges the intrinsic characteristics of a wave-corpuscle electron, which are (a) the spatial extension of a corpuscle electron exhibits as a delta function that is usually expressed in the coordinate space

$$\delta(r) = 1/(2\pi\alpha)^{1/2} \int \exp(-r^2/2\alpha)dr$$

which is a Gaussian distribution function with $\Psi_0 = \kappa_0 A_0 B_0 = (4\pi^2 R_0 r_0)^{-1/2}$. This means the spatial extension of a wave-corpuscle electron is related to the radius of the rotating and circulating photon. (b) The dynamic periodic rotating process is expressed as

$$\exp(ikz)\exp(i\mu\omega t)$$

which is the spatial and time periodic function in its own frame. This clearly indicates that in the laboratory frame, the twisted circulating photon as a vorticity field is rotating around the center of the torus while it is at rest. When it moves with the velocity v_t, the corpuscle electron frame moves with a velocity v_t corresponding to the laboratory frame. The Lorentz transformation has to be applied to the physical parameters in both frames. The Lorentz transformation gives us a description of space-time in which the notions of simultaneity, time duration, and spatial distance are well defined in each inertial reference frame, but physical parameter values for a given pair of events in an inertial moving frame can be different from the laboratory frame. In particular, intervals of space and time in frames moving with relative velocity v_t are related to the Lorentz transformations

$$z' = \gamma[z - (v_t/c)ct], ct' = \gamma[ct - (v_t/c)z]$$

or

$$z = \gamma[z' + (v_t/c)ct'], ct = \gamma[ct' + (v_t/c)z'].$$

If we choose z and t in the laboratory frame and z′ and t′ in the corpuscle electron frame moving with velocity v_t, the frequency and the interval of the tracks of the rotating vorticity photon would be different between the laboratory and corpuscle frames.

As previously discussed, in the laboratory frame, we see the electron as a charged corpuscle with mass. However, in reality, the corpuscle electron, as a topological torus, has its own inertial reference frame. A corpuscle electron, in fact, is nested in a twisted circulating (vorticity) photon rotating along the curving surface of a torus. In other words, an electron is a tiny space as a corpuscle, in which a periodic dynamic motion of electromagnetic energy flux, creating the charge and mass of the electron, is confined. Usually, velocity v_t of an electron is the velocity of the corpuscle in the inertial laboratory frame, but not the photon's velocity. Therefore, the wave function, Ψ (r, θ,φ) = Ψ_0 [δ(r-ρ) exp (ikz) exp (i$\mu\omega$t)] = ψ(r)exp [i(kz-ωt)], has to be elucidated by the electrodynamics' process, but not as a probabilistic plane wave.

First, assuming an electron moves in one direction, z, with velocity v_t, and if the time t_0 is the time in the corpuscle frame and t is the time in the laboratory frame, then

$$t_0 = 1/\gamma\left(t - v_t z/c^2\right) \text{ where } \gamma = \left(1 - (v_t/c)^2\right)^{1/2}$$

and $\omega t_0 = (\omega/\gamma)\left(t - v_t z/c^2\right)$, so $\omega/\gamma = \omega_w$, which implies the rotation frequency of the twisted circulating photon is increased. Therefore, the periodic time for the track of the photon traveling on the surface of a torus is getting shorter, resulting in the velocity of the track, which is the phase velocity $v_{ph} = c^2/v$ at the electron's internal frame, being faster than the speed of light. It should be emphasized that the kinetic motion of an electron does not destroy the internal dynamic process of the twisted circulating photon, but modifies the kinetic parameters or characteristics (e.g., frequency, phase velocity, mass, etc.) of periodic dynamic motion in an electron's internal frame. This feature is posted on the $\Psi_\varphi = B_0 \exp(i\mu\varphi) = B_0 \exp(i\mu\omega t)$ function.

Second, because of the electron's motion, the rotation frequency has to increase in the moving direction as mentioned earlier,

$$\omega_t = \omega/\gamma \approx \omega\left(1 + 1/2(v/c)^2\right) = \omega + 1/h\left[(1/2)\ m_0 v^2\right] = \omega + \omega_v$$

This relationship indicates that the rotating frequency of the vorticity photon around the center of the torus in the internal frame has gained an additional rotation frequency, which emanates from the moving velocity v_t of the corpuscle or the torus. Although the motion velocity of the corpuscle electron may be linear speed, the dynamic process of the twisted circulating photon in its own frame is still circulation and rotation, and the rotation frequency is increased in the moving direction. Understanding this phenomenon is important for understanding the duality of the particle and wave nature of the electron or de Broglie's idea about matter waves. The frequency of de Broglie's matter wave corresponds to the increased kinetic energy of the confined electromagnetic energy in the corpuscle electron.

The ratio of ω/ω_v shows the Doppler effect, which is

$$\omega_v/\omega = \left[(1 + v_t/c)/(1 - v_t/c)\right]^{1/2}$$

It is worth remembering that velocity is a vector that can be decomposed into longitudinal and transversal directions in space. For the corpuscle electron frame, the basic unit vectors are τ_φ, τ_θ, and τ_n as shown in Figure 1.10, and the modified angular frequencies ω have different values in different orientations. If the z-axis of the torus is in the moving direction as the longitudinal direction, the traversal Doppler effects for a corpuscle electron is

$$\omega_v/\omega = \gamma/(1-(v_t/c)\cos\varphi)$$

This effect leads to modifying $\zeta = r_0/R_0$ of the torus, making the spatial extension of the torus smaller than the radius of the circulating photon and larger than the radius of the rotating circle of the photon's internal frame. However, in the momentum space, the momentum of the photon in the propagation direction increases and the transverse momentum of the photon decreases. The spatial extension of the corpuscle electron will change from nearly spherical to elliptical, as shown in Figure 1.13.

Note that (1) an electron is a confined electromagnetic energy flux in a tiny space, and the tracks of the electromagnetic energy flux comprise an assembly of the tracks of the moving photon in a periodic time and a periodic path on the surface of a torus. These two periodic motions of the tracks of the moving photon are synchronized to form periodic electromagnetic wave fields. This can be seen as a monochromatic electromagnetic harmonic oscillator creating a confined electromagnetic energy entity. The electromagnetic energy flux is the Poynting vector $P = 1/c(E\text{x}B)$.

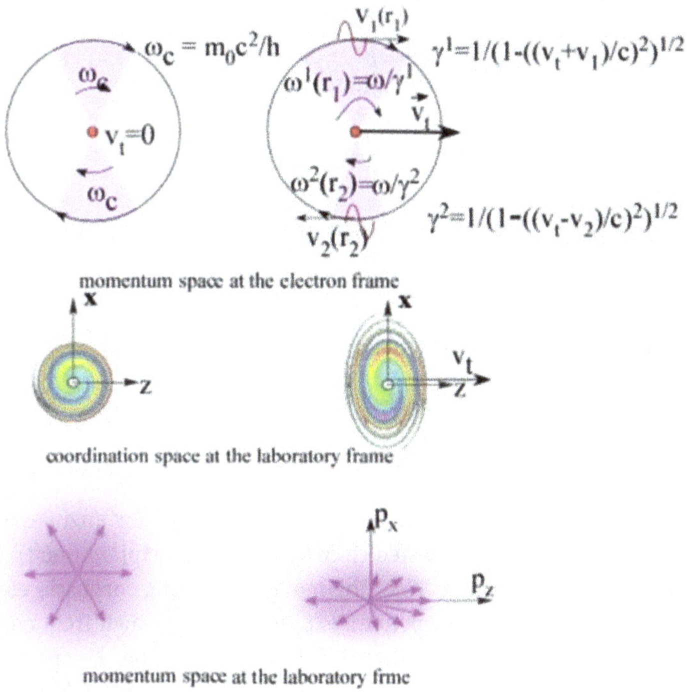

Figure 1.13: The moving electron with velocity v_t observed in the laboratory frame will modify the kinetic process of the twisted circulating photon in the electron frame by the Doppler effect. This causes a change in the morphology of the electron in the configuration space and momentum space.

(2) The properties of an electron such as mass, charge, and spin are the results of the kinetic curved movement process of the electromagnetic energy flux along the two perpendicular planes. This movement obeys Newton's classic dynamic law and extended Maxwell's equation. Therefore, the energy variation with time for the circulating photon can be expressed as $i\hbar(\delta\mathcal{E}/\delta t)$ in the laboratory frame.

(3) The electron is a real entity, such as the twisted circulating photon rotating around the center of the corpuscle electron, and the entity is a real object with mass, charge, and spins, similar to the matter seen in our daily life. The function Ψ (r, θ, φ) (usually called the "wave function") carries the information about these real objective fields.

1.3.4 Wave function, Maxwell equation, and Schrödinger equation

The wave function Ψ (r, θ, φ) expresses the dynamic process of the vorticity photon in the corpuscle electron. To describe a moving electron, we have to distinguish between the dynamic process of the internal process at rest and the kinetic process in the laboratory frame. For a general view, ω relates to $\mathcal{E}$ – energy and z is similar to r in torus model, and so the function would be $\Psi(r,\theta,\varphi) = \Psi_0 \exp((1/h)(pr - \varepsilon t))$.

If using energy equation, $\varepsilon^2 = p^2c^2 + m_0{}^2c^4$ and

$$\begin{aligned}\varepsilon &= mc^2\sqrt{1+\frac{p^2}{m^2c^2}},\\ &\approx mc^2\left(1+\frac{1}{2}\frac{p^2}{m^2c^2}\right),\\ &\approx mc^2+\frac{p^2}{2m} = mc^2+\mathcal{T}\end{aligned}$$

$\mathcal{T}$

The wave function can also be represented as

$$\begin{aligned}\Psi(x,t) &= \Psi_0 e^{\frac{i}{\hbar}(px - mc^2t - \mathcal{T}t)}\\ &= e^{-\frac{i}{\hbar}mc^2}\Psi_0 e^{\frac{i}{\hbar}(px-\mathcal{T}t)}\end{aligned}$$

This wave function has two components: one is the intrinsic energy (or angular frequency) of the internal oscillation of the twisted circulating photon, $\exp[-i/\hbar(mc^2t)]$ or $\exp[-i\omega t]$ (~10^{22} s^{-1}), and the energy gain, $\mathcal{T}$, of the movement of the corpuscle electron in the laboratory frame. The first component, $\exp[-i\omega t]$, is super-high frequency, but the other is a lower oscillation, except that the velocity would be the speed of light c, and the frequency changes with the velocity of an electron in the laboratory frame. The momentum, p, and coordination position, r, of the traveling corpuscle electron are expressed in the laboratory frame; therefore, we can rewrite the wave function as

$$\Psi(r,\theta,\varphi)=\Psi_v\exp\left(-(i/\hbar)m_0c^2t\right),$$

where $\Psi_v=\Psi_0\exp[i/\hbar(pr-\tau t)]$or $\Psi_v=\Psi_0\ \exp[i/\hbar\ (pr-\varepsilon_v t)]$. Therefore, the general wave function of a moving electron is

$$\Psi(r,\theta,\varphi)=\Psi_v\exp\left(-(i/\hbar)m_0c^2t\right)$$

However, the commonly used function is

$$\Psi_v=\Psi_0\exp[i/\hbar(pr-\varepsilon_v t)]$$

This function does not demonstrate the internal structure of an electron; it only describes the movement of the electron in the laboratory frame. It is important to note that pr/ħ is the intrinsic ratio, because pr/h is related to the mass and the frequency of the internal rotation of the electron by using Planck's constant. As pr approaches ħ, the internal structure of the electron must be affected.

Putting the wave function into the general wave equation, then we can obtain

$$e^{-\frac{i}{\hbar}mc^2t}\left[\frac{\partial^2}{\partial x^2}+\frac{2im}{\hbar}\frac{\partial}{\partial t}\right]\phi=0\,\frac{\partial^2\phi}{\partial x^2}+\frac{2im}{\hbar}\frac{\partial\phi}{\partial t}=0$$

Then the final equation is

$$-\frac{\hbar^2}{2m}\nabla^2\phi=i\hbar\frac{\partial\phi}{\partial t}$$

This is what usually uses the Schrödinger equation. We may emphasize that (1) ϕ is the distribution function of the electric field of the electron; (2) the ϕ function is the wave function without intuitively demonstrating the internal circulation of the vorticity photon on the torus, but it is related to the internal movement of the vorticity photon. For example, if the electron interacts with a higher frequency electromagnetic field, the internal circulating energy of the vorticity photon has to be considered, such as in the Kapitza-Dirac effect. In general, for electron microscopy, it need not be considered. Recently, Howie using a standing wave of strong laser waves with proper frequencies to interact with electrons to disclose the phase information during the interaction between fast electrons and the sample to be observed.

We may use this wave function in momentum space to find the wave function in space in the laboratory frame:

$$\Psi(x,t)=\frac{1}{\hbar\sqrt{2\pi}}\int_{-\infty}^{+\infty}\phi(p,0)e^{i(p/\hbar)x-i(E/\hbar)t}dp$$

$$\phi(p,0)=\frac{1}{\sqrt{2\pi}}\int_{-\infty}^{+\infty}\Psi(x,0)e^{-ipx/\hbar}dx$$

It is easier to understand this function as $\phi(p, t)$. If we see an electron as a particle without considering the internal nature of the vorticity photon, then the momentum of the electron should have an original value p_0 corresponding to the internal circulation of the vorticity photon at rest. When it is freely moving with velocity v_t, the momentum of the rotating vorticity photon varies around the original p_0 value. Then it may use a delta function to describe the function in the momentum space.

$$\Phi(p,0) = A\delta(p - p_0)$$

where A is some constant. The wave function, which is a solution to the Schrödinger equation,

$$\psi(x,t) = \frac{1}{\sqrt{2\pi\hbar}} \int_{-\infty}^{+\infty} A\delta(p - p_o) e^{i(p/\hbar)x - i(E/\hbar)t} dp$$

or

$$\psi(x,t) = \frac{A}{\sqrt{2\pi\hbar}} e^{i(p_o/\hbar)x - i(E_o/\hbar)t}$$

where $\mathcal{E}_0 = p_0^2/2\,m$ for a free electron. However, this function shows a plane wave and a constant density everywhere. It is not localized. Therefore, the plane wave function cannot describe the electron. The function $\phi(p, t)$ cannot be a delta function; in other words, the wave function of the electron has to relate to the internal nature of the electron. Based on the vorticity photon model of electron, momenta of the photon have to distribute as a vorticity photon traveling on the torus or some topological functions such as a sphere, cylinder, or cone. For a free electron, these functions may be a Gaussian distribution or Green's function. If we consider the confined electromagnetic waves in a corpuscle electron, it is better to use the Dirac delta function in momentum or wave vector space:

$$\delta(k - k') = \frac{1}{2\pi} \int_{-\infty}^{+\infty} e^{i(k - k')x} dx$$

which is equivalent to

$$\delta(p - p') = \frac{1}{2\pi\hbar} \int_{-\infty}^{+\infty} e^{i(p - p')x/\hbar} dx$$

This means that the expectation value of the momentum can be written as

$$\langle p \rangle = \int_{-\infty}^{+\infty} dp\phi^*(p,t) p\phi(p,t)$$

Therefore, the momentum may vary widely; however overall, in the whole momentum space, it may have a definite average value as $p_0 = \mathcal{E}_0/c = m_0c$.

We may have

$$\langle x\rangle = \int_{-\infty}^{+\infty} dp\phi^*(p,t)\left(i\hbar\frac{\partial}{\partial p}\right)\phi(p,t)$$

This indicates that the position expectation value $<x>$ may be obtained from the distribution function of the momentum of the electron. This expectation value gives the localized position of the corpuscle electron, or the particle-like electron. Because the electron has its own frame, the expectation values never have a definite unique value in the laboratory frame; instead they have some range Δx. If we use the following function as the wave function of a corpuscle electron

$$\psi(x,0) = \begin{cases} \frac{1}{\sqrt{\Delta x}} & \text{if} |x| < \frac{\Delta x}{2} \\ 0 & \text{if} |x| > \frac{\Delta x}{2} \end{cases}$$

Fourier transformation of this function is as follows:

$$\phi(p,0) = \frac{1}{\sqrt{2\pi\hbar}}\int_{-\infty}^{+\infty} \psi(x,0)e^{-ipx/\hbar}dx$$

This may give the momentum distribution function of the traveling corpuscle electron as

$$\phi(p,0) = \frac{(2\hbar/p)}{\sqrt{2\pi\hbar\Delta x}}\sin(p\Delta x/2\hbar)$$

It is valuable to indicate that if Δx decreases when $p\Delta x/2\hbar = n\pi$ or where $p = n2\pi\hbar/\Delta x$, the value of p, where $\phi(p, 0) = 0$, increases. If the uncertainty in the p domain is defined as the width of the function $\phi(p,0)$, i.e., the distance between the first and second zeros of the function, then we will have $\Delta p = 4\pi\hbar/\Delta x$ or $(\Delta p)(\Delta x) = 4\pi\hbar = 2h$. Thus, as $\psi(x,0)$ becomes more localized (or Δx is decreases), then $\phi(p,0)$ will be more spread out. This is the Heisenberg uncertainty principle, which may originate from the intrinsic nature of the corpuscle electron.

1.3.5 Relationship between the density of mass and charge and the wave function

The mass and charge arise from the rotation and circulation of a photon on the surface of a torus in a time period t_0. The density of the mass and charge is $\rho_m = m_0/4\pi^2 Rr$ and $\rho_e = q_e/4\pi^2 Rr$. Because $\Psi\ (r,\theta,\varphi) \bullet \Psi^*\ (r,\theta,\varphi) = |\ \Psi_0(r,\theta,\varphi)|^2$ should be 1, a confined electromagnetic energy should exist in this tiny space. The mass and energy density of an electron should be equal to

$$\rho_m = m^2/4\pi 2R_0r_0 = (h\omega/c^2)^2/4\pi 2R_0r_0$$

and

$$\rho_{em} = (h\omega)^2/4\pi 2R_0r_0$$

Then,

$$|\Psi_0(r,\theta,\varphi)|^2 = \rho_m \text{ or } |\Psi_0(r,\theta,\varphi)|^2 = \rho_{em}.$$

For an electron, as a moving corpuscle in the laboratory frame, shown in Figure 1.14, the relationship between the coordination of the starting element volume, dV(t), and the transported element volume, dV(t + dt), should be expressed as

$$dV(t_0 + dt) = J(t_0 + dt, t_0)dV(t_0)$$

where J (t_0 + dt, t_0) is Jacobian transformation.

Jacobian transformation is given by

$$J(t, t_0) = \exp\left[\int_0^t (\nabla \bullet v_t)dt\right]$$

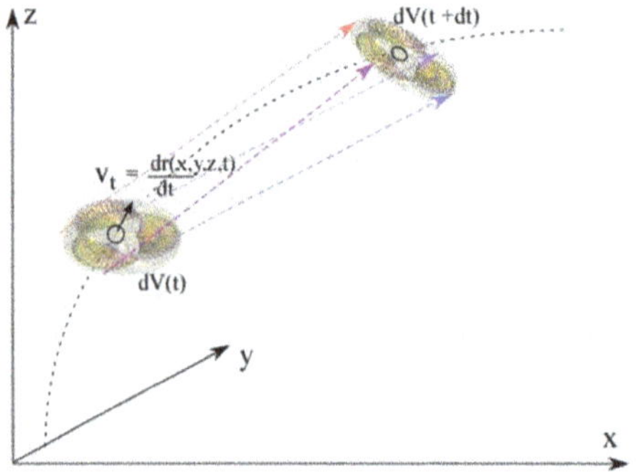

Figure 1.14: Area elements at the start time t_0 and the transported time t_0 + dt and the ratio between them is the Jacobian transformation.

Jacobian transformation is the solution of the following differential equation:

$$dJ/dt = (\nabla \bullet v_t)J$$

This equation is the same as the Lagrangian version of the continuity equation for the probability density. The initial condition on the solution is J (t_0, t_0) = 1 and the ($\nabla \bullet v_t$) is called the “compressibility of the system”; it may be used as a measurement of the degree of expansion and contraction of a corpuscle electron on the trajectories.

These trajectories may describe an assembly of the tracks of rotation and circulating photons or the swarm of many paths of the vorticity photon, without considering the internal motion of the vorticity photon. In general, the coordination of a corpuscle electron in the laboratory frame is the spatial position and the velocity dr/dt means the speed of the mass center of the corpuscle electron. The volume of the corpuscle electron is spatial volume occupied by the rotating vorticity photon. The spatial volume is incompressible except for changes in energy of the electron.

As mentioned, if the mass and charge density is

$$\Psi_0(r,\theta,\varphi) = \sqrt{\rho_m} \text{ or } \Psi_0(r,\theta,\varphi) = \sqrt{\rho_{em}}$$

then the wave function would be

$$\Psi_v(x,y,z,t) = \sqrt{\rho}\, \exp\left[i/\hbar(p(x,y,z,t)r(x,y,z,t) - \varepsilon_v t)\right]$$

When we discuss the moving electron in the laboratory frame, the equation of dynamic motion has to be used. The particle's dynamic motion can be described by the Lagrangian theory, but the wave's dynamic motion would use the Eulerian space. They are linked to each other by the Eulerian density:

$$\rho(x,t) = \int \delta(x - q(a,t))\rho_0(a)d^3a$$

This relationship can be regarded as components of transformation from the coordinate functions, $q_i(a)$, to a new system of coordinates as $\rho(x)$, which is a building block of the canonical transformation linking the two pictures, i.e., particle and wave. The corresponding formula for Eulerian velocity v is contained in the expression for the current as

$$\rho(x,t)\; v_i(x,t) = \int (\delta q_i(a,t)/\delta t)\delta(x - q(a,t))\rho_0(a)d^3a$$

These formulas are analogues of the Huygens formula. The delta function $\delta(x\text{-}q(a, t))$ is the propagator of the motion in time. Because

$$\delta(x - q(a,t)) = J^{-1}{}_{a(x,t)}\delta(a - a_0(x,t), x - q(a_0,t) = 0$$

$$\rho(x,t) = J^{-1}{}_{a(x,t)}\rho_0(a(x,t)), \quad \rho(x(a,t),t) = J^{-1}(a,t)\rho_0(a)$$

The formulas tell us that if the initial distributions of density are $\rho_0(a)$, the density at any given time can be obtained by an anti-Jacobian transformation, or if the initial density at a position is known, the density at any position and at any time can be obtained by an anti-Jacobian transformation. The particle view for an electron can be transformed into a wave picture.

If we substitute the wave function of a corpuscle electron

$$\Psi_v(x,y,z,t) = \sqrt{\rho}\, \exp[i/\hbar(p(x,y,z,t)r(x,y,z,t) - \varepsilon_v t)]$$

into the Schrödinger equation,

$$i\hbar\frac{\partial\psi(x,t)}{\partial t} = -\frac{\hbar^2}{2m}\frac{\partial^2\psi(x,t)}{\partial x^2} + V(x,t)\psi(x,t)$$

Using its complex conjugate function to multiply both sides, we may have

$$\frac{\partial|\psi(x,t)|^2}{\partial t} = i\frac{\hbar}{2m}\frac{\partial}{\partial x}\left(\psi^*(x,t)\frac{\partial\psi(x,t)}{\partial x} - \psi(x,t)\frac{\partial\psi^*(x,t)}{\partial x}\right)$$

Because $|\Psi_v|^2 = \rho(x, y, z)$, and $v_t\rho(x, y, z) = J$, is the mass or charge current of the corpuscle electron, that is,

$$J(x,t) = i\frac{\hbar}{2m}\left(\psi(x,t)\frac{\partial\psi^*(x,t)}{\partial x} - \psi^*(x,t)\frac{\partial\psi(x,t)}{\partial x}\right)$$

the velocity, v, is

$$\upsilon(x,t) = \frac{J(x,t)}{|\psi(x,t)|^2}$$

This means the Schrödinger equation describing the wave picture of corpuscle electron is the current equation of the corpuscle "particle" in the Lagrangian system. The velocity is the speed of the mass center of the internal frame of a corpuscle electron.

We used $\Psi(r,\theta,\varphi) = \Psi_v \exp(-(i/\hbar)m_0c^2t)$ to express the dynamic process of the twisted circulating photon; now we can use another way to describe this process, which consists of two wave functions, $\Psi = \Psi_R(x,t) + i\Psi_\theta(x,t)$ and uses the polar coordinates

$$|\Psi|^2 = |\Psi_R|^2 + |\Psi_\theta|^2 = \rho_r{}^2 + \rho_\theta{}^2 = \rho^2 \text{ and } S(x,t) = \hbar\arctan(\Psi_\theta/\Psi_R)$$

S(x,t) is the so-called quantum action function; if $\Psi_R = \Psi_\theta = 0$ that implies that $\rho(x, y, z) = 0$. There is no particle. If we use $\Psi = \rho(x,t)\exp(iS(x,t)/\hbar)$ as a wave function and substitute it into the Schrödinger equation, then

$$\frac{\partial R^2(x,t)}{\partial t} + \frac{\partial}{\partial x}\left(\frac{1}{m}\frac{\partial S(x,t)}{\partial x}R^2(x,t)\right) = 0$$

where $R = \rho$. This is the local conservation law. Then

$$\frac{\partial S(x,t)}{\partial t} + \frac{1}{2m}\left(\frac{\partial S(x,t)}{\partial x}\right)^2 + V(x,t) + Q(x,t) = 0$$

This is the quantum Hamilton-Jacobi equation. Q(x,t) is called the "quantum potential," which is defined as

$$Q(x,t) = -\frac{\hbar^2}{2m}\frac{\partial^2 R(x,t)/\partial x^2}{R(x,t)}$$

Therefore, we have an interpretation of the wave function, which is the solution to the Schrödinger equation as an assembly of quantum trajectories with different initial positions and velocities. The velocity at each trajectory x(t) is defined as

$$\upsilon[t] = \left[\frac{1}{m}\frac{\partial S(x,t)}{\partial x}\right]_{x=x[t]} \qquad \upsilon[x,t] = \frac{1}{m}\frac{\partial S(x,t)}{\partial x} = \frac{J(x,t)}{|\psi(x,t)|^2}$$

From these equations, we can easily find a Newton-like equation such as the following:

$$m\frac{d}{dt}\upsilon(x[t],t) = \left[\frac{\partial}{\partial x}\left(\frac{1}{2m}\left(\frac{\partial S}{\partial x}\right)^2 + \frac{\partial S}{\partial t}\right)\right]_{x=x[t]}$$

or

$$m\frac{d}{dt}\upsilon(x[t],t) = \left[-\frac{\partial}{\partial x}(V(x,t) + Q(x,t))\right]_{x=x[t]}$$

Based on this, we may find a common language for classical and quantum theories. The quantum complex single-particle wave function can be interpreted as an assembly of trajectories that are all solutions of quantum, Newton-like equations for the same single particles but with different initial conditions. The quantum trajectories are not solutions of Newton's classical second law with a classical potential, but a solution of Newton's quantum second law, in which a quantum potential that accounts for all non-classical effects is added to the classical potential.

The differences between quantum and classical assemblies of trajectories are not differences between waves and particles, because both waves and particles can be used to study classical or quantum systems. The differences are rooted in a linear wave equation for quantum mechanics and a nonlinear wave equation for classical mechanics. One of the most important consequences of such differences is that quantum trajectories depend on the shape of the assembly, that is, the topologically confined rotation space of vorticity photon on the surface of the torus, whereas classical trajectories are independent of the shape of the assembly.

In electron microscopy at low and medium magnification ($<10^5$), the electron can be seen as a corpuscle without considering the quantum potential or the internal dynamic process of the electron, but the interaction between the electron and the sample, and the influence of a magnetic lens on the electron have to considered using the quantum potential rigorously.

We may summarize the essential difference between a particle and a wave as follows.

A classical particle follows a trajectory in space as time flows and is associated with a classical Hamiltonian, which has three spatial degrees of freedom. It exhibits a fixed mass and an internal structure, and its interactions follow the conservation of energy, momentum, and angular momentum. One type of trajectory is a periodic oscillation or rotation, and the amplitude of this oscillation is unrestricted. Many interacting particles may form an ensemble, the motion of which can be decoupled into collective modes, being largely independent. A particle generally has a defined size with a center of mass.

A classical wave is a time-varying spatially extended field, a real distributed oscillation in real space, generally described by a linear wave equation, which can be analyzed in terms of wave components with ω and k: the wave components can be subjected to linear superposition. Such a real wave may be modeled by a complex wave function for mathematical convenience, such as

$$\psi_{cl}(x,t) = R(x,t)\exp(iS(x,t)/\hbar)$$

A wave does not have a fixed size, but may be confined by boundaries to a region that is larger than half a wavelength. The confined wave forms a discrete but infinite set of standing wave modes, each having its own value of ω and k, similar to classical electromagnetic waves. Noninteracting waves can generally share space and pass through each other. There are scalar and vector waves, which may have polarization. Waves can be coherent or incoherent in space and time; coherent waves are associated with effects such as interference and diffraction.

A true quantum wave has quantized mode frequencies, and the amplitudes are also quantized. This gives rise to the quantization of angular momentum (spin), linear momentum, and energy. A quantum wave is fundamentally a complex wave exp(iϕ), rather than a real oscillation in real space, but this is really an artifact of the mathematical model rather than a fundamental aspect. In fact, the real oscillation frequency of a quantum wave is given by its full relativistic energy ($\omega = mc^2/h$), but this frequency is generally offset in the nonrelativistic case. In terms of standing waves and superposition, a quantum wave behaves much like a classical one. Transitions between waves of differing quantized amplitudes have a distinct quantum character.

The "quantum particle" follows a classical trajectory but may also have a localized coherent phase and quantized amplitude originating from the internal dynamic motion of the electromagnetic wave flux. A confined quantum wave (such as an electron) in its motion acts as a quantum particle in the laboratory frame.

1.3.6 Properties of a moving electron with velocity υ(t) in the laboratory frame

The moving electron with velocity υ(t) in the laboratory frame may be described as the trajectory of a corpuscle electron with a small volume as ξ(x, y, z, t) and υ(t) = d ξ/dt. The trajectories may come from the velocity at each position in space over time by integration with a time period as $\xi(x, y, z, t) = \int \upsilon(t)\, dt + \xi_0$. The trajectory is the path of the center of a corpuscle electron in its own frame, in which there are two frames having two-dimensional circles that are perpendicular to each other. One of these frames is parallel to the laboratory frame; the other is perpendicular to this plane. The real actor, such as a monochromatic photon with energy $\mathcal{E} = m_0c^2 = h\omega$, moves with a twisted circulating motion and simultaneously rotates around the center of the corpuscle or torus. Therefore, the tracks of this actor make the trajectory of the corpuscle electron. However, the moving trajectory of this actor is a synthesis of two configuration manifolds $(S^1)^2$, in which an inertial, fixed Euclidean frame is constructed so that the center of the circle is located at (R,0,0); the circle lies in the x-z plane, with the z-axis as the rotation axis of the vorticity photon. The position of the vorticity photon actor may be expressed as x = (x_1, x_2, x_3) in the laboratory frame, and the tracks' configuration may be selected as q = (q_1, q_2), which uniquely determines the position vectors of tracks on the surface of a torus. Therefore, the photon only has two degrees of freedom as shown in Figure 1.10. The description of corpuscle electron movement actually has two equally exchangeable pictures, such as Lagrangian and Euclidean pictures, as Figure 1.15 shows.

The wave function describing this corpuscle electron may be expressed as a synthesis of three parts: (a) a Jacobian transformation between the coordinates in the laboratory frame and a set of coordinate of the tracks of the vorticity photon in the corpuscle frame, which comprise two perpendicular two-dimensional systems; (b) the motion of the tracks assembled of circulating photons during the rotation periodicity, which may be described as the oscillation of the track of the photon or the electric field's intensity oscillation in the torus; or (c) the motion of the whole corpuscle described as a particle traveling along a trajectory ruled by the Lagrangian principle, in which the trajectories are dominated by the action force or potential gradient at each point on the trajectory. Therefore, the wave function is time-dependent as

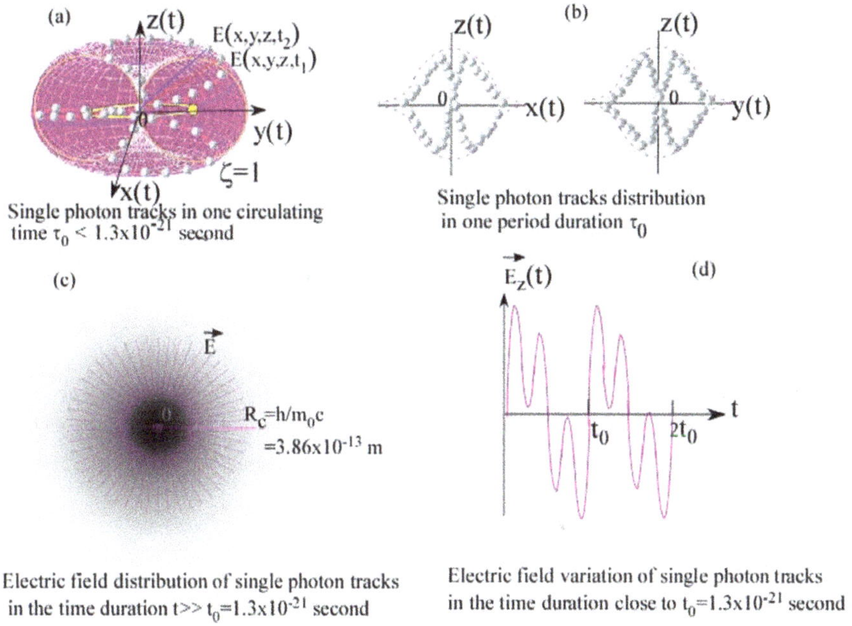

Figure 1.15: (a) The tracks of a moving photon on the surface of a torus in one periodic time τ_0, (b) the tracks of a moving photon projected into the x-z and y-z planes in one periodic time τ_0, (c) the electric field created by the moving photon in the laboratory frame as a charged corpuscle, and (d) the variation of the electric field intensity in the configuration space of the torus.

$$\Psi(r,t) = J(t)^{-1/2}\exp\{i/\hbar \left[\int_{t_0}^{t} L(t)dt\right] \Psi(r_0, t_0)$$

and

$$\rho(r,t)J(t,t_0) = \rho(r_0,t_0)$$

which expresses conservation of $\rho(r,t)\ J(t,t_0)$ along the trajectory.

$J(t)^{-1/2} = \exp\{\int \nabla \cdot vdt\}$(integrating from t_0 to t) is the divergence of the velocity field, and if it is positive, it indicates that the volume of the corpuscle electron will increase along the flow. So this $J(t)^{-1/2}$ describes the divergence of the velocity of the corpuscle electron; $(\nabla \cdot v_t) = 0$ means the Jacobian is invariant and the flow of the corpuscle electron is incompressible. Two corpuscle electrons cannot arrive at the same position and at the same time; experimental results have shown as the anti-bunching properties of electrons called the "Hanbury Brown and Twiss (HBT) effect." This also implies that the configuration shape of the corpuscle electron may be modified, but its volume is constant along its trajectory. If the electron gains some energy from its environment or loses energy due to the radiation, based on the torus model, $\zeta = r/R$ will be changed and the configuration morphology should be modified. The core of the oscillation energy of the vorticity photon as the energy carrier, which is $\mathcal{E}_0 = m_0c^2 = h\omega_0$, does not change, but the kinetic and potential energy gain of the electron will cause the frequency of rotation of the vorticity photon to change. In other words, in general, the increase in the velocity of an electron may vary the periodic paths of rotation of the vorticity photon but not change the vorticity motion of the photon significantly, unless the velocity is close to the light speed c where the photon would propagate linearly without any curvature as a γ-ray.

In summary, an electron is a wave-corpuscle that emerges from the dynamic circulation process of the electromagnetic energy flux of a photon confined in two perpendicular dimension circles forming a configuration manifold $(S^1)^2$ as a torus. The diameter of the corpuscle electron is the Compton wavelength $\lambda_c = h/m_0c = 3.86 \times 10^{-13}$ m, and the mass and charge come from a photon's twisted circulation in one circle and rotating around another circle that is perpendicular to the previous one. The Poynting vector at each point on the surface of a torus can be decomposed into two perpendicular tangent velocities, v_{t1} and v_{t2} on the circumference of the two circles. The circulation integral

$$I = \int v \cdot dl = (1/m) \int pdl = (1/m) \int \nabla S \cdot dl = (1/m)\ \Delta S$$

is the change in action from one periodic transition around the circle. For the twisted circulating loop of the photon, $I_\theta = \mathcal{E}_0/\omega = h$ (because $\mathcal{E}_0 = \omega\ I_\theta$). For the rotational motion,

$$I_\varphi = \int v \cdot dl = \int L_\varphi d\varphi = L_\varphi = mR^2\omega$$

is the angular momentum of the corpuscle electron. This tells us that the electromagnetic energy of an electron resides in the twisted circulating motion of a photon with a super oscillation electromagnetic field, $\omega \approx 10^{22}$ Hz. The external electric field of the corpuscle electron may be created as the time-averaged electric field, which is the charge field of the corpuscle electron (1.6×10^{-19} C). These motions introduce variations in the angular momentum emanating from changes in the velocity orientation of the corpuscle electron along its trajectory. The energy gain or loss of a corpuscle electron by absorbing or emitting radiation will modify these motions but never destroy the twisted circulating dynamic process unless the velocity of the corpuscle reaches the speed of light, c. The scattering process cannot detect the internal structure of a corpuscle electron.

The initial wave function is very important for the determination of the state at any position of trajectory along the traveling paths. That is why the emitted electron at the electron gun is important for electron microscopy. The paths of the corpuscle electron can be found by quantum dynamic mechanics, especially with dynamic motion equations such as the Schrödinger equation, the quantum Hamilton-Jacobi equation, the Maxwell equation, and so on. These equations seem different, but they are equivalent. All of them exhibit the nature of energy flow under different kinetic conditions, such as in linear, curved, spiral, or vortex motion.

In an electron microscope, the state of an electron at its starting position, near the surface of the filament of an electron gun, may be seen as nearly at rest, with m_0. The electrons absorb energy from the accelerating electric field; they increase their velocity and fly through some area (called the image of the filament of the gun) at different times. The kinetic energy will be $\mathcal{E}_K = (1/2)\,(m_0/\gamma)\,\upsilon^2$. Electron microscopy may be seen as the measurement or encoding of information on the velocity modification while the electron travels from the starting state at the imaged place of the filament of the gun to the final state at the recording device. The electron as a corpuscle has a topological structure, and the electromagnetic energy flux has two types of the motion. Therefore, the detectable motion is the kinetic motion of the corpuscle electron, such as spatial interval variation, which is wavelength, and frequency modification of the angular momentum, which is the energy variation of the twisted circulating photon. The anti-bunching effect of the electron is a result of the internal structure of the corpuscle electron. This distinguishes the characteristics of the electron from those of the photon.

1.4 de Broglie wave and internal oscillation of electromagnetic energy flux of the corpuscle electron

Particle nature of the electron was the idea behind Ruska's invention in 1931, but imaging obtained by a transmission electron microscope shows that the electron is both a particle and a wave as previously discussed. To understand the coded information on an image photograph from a transmission electron microscope, it is essential to understand the nature of the electron. Advanced elementary particle physics and quantum mechanics, especially the progress of de Broglie-Bohm mechanics, reveals the nature of de Broglie matter wave as confined electromagnetic energy, as mentioned previously. Now we may comprehensively understand the de Broglie matter wave and ponder what electron microscopy is.

1.4.1 What is the nature of the de Broglie wave?

In 1924, based on combining Einstein's mass energy equivalence, $\mathcal{E} = mc^2$, with the Planck-Einstein law $\mathcal{E} = h\nu$, de Broglie proposed that the wave characteristics of matter, $p = h/\lambda$, come from internal oscillation with a fixed frequency, which is equal to the frequency of a photon, and its energy is equal to the rest energy of the matter particle. The internal oscillation creates a phase wave in a moving particle; the kinetic energy of a freely moving electron in the laboratory frame is $\mathcal{E}_k = ½\ mv^2$ or $\mathcal{E}_k = (1/2\ m)\ p^2$, in which m is inert mass and v is the velocity observed in the laboratory frame. However, the mass of the electron emanates from the velocity change of electromagnetic energy flux of a photon circulating at the speed of light in its internal frame. Therefore, an observer in the laboratory frame would only see the moving electron as a corpuscle with speed v, but the kinetic process parameters in the corpuscle have to use Lorenz-transformed values due to the relative motion of two inert frames, such as

$$\delta m = mv\delta v/c^2\left(1 - (v/c)^2\right);$$

For Newtonian mechanics, the kinetic energy can be expressed as

$$\delta\varepsilon = fv\delta t = \delta[(mv)v] = v^2\delta m + mv\delta v,\ \text{and}\ \delta\varepsilon = c^2\delta m.$$

If we integrate from $v_1 = 0$ to $v_2 = v$, the kinetic energy is $\varepsilon_k = mc^2 - m_0c^2$. From mass transformation $m = m_0/(1 - (v/c)^2)^{1/2}$, we may find the relationship between energy and momentum of the moving corpuscle electron in the laboratory and corpuscle frames to be $\varepsilon_e{}^2 - c^2p_e{}^2 = \varepsilon^2 - c^2p^2 = m_0{}^2c^4$, respectively. It tells us that the total energy of the electron at rest is different from the energy of momentum. In other words, the total energy observed in the laboratory frame has two components: kinetic energy coming from the corpuscle electron's motion, and intrinsic energy emanating from the dynamic internal process of the circulating photon. Because the energy of the corpuscle electron is $\varepsilon^2 = m^2c^4$, it is easy to derive the equation,$m = m_0 + p^2/c^2$. Using $p = h/\lambda$, $m_0 = h\omega_0/c^2$, and $p = mc\beta(\beta = v/c)$, we can obtain $\omega^2 = \omega_0{}^2 + \omega_{db}{}^2$, which divulges the de Broglie wave phenomenon as a relativistic effect of the internal rotating electromagnetic wave of vorticity photon due to length contraction and time dilation. This gives rise to the vector mode of oscillation of the electromagnetic field energy flux of the photon. The group speed of these oscillation waves is the velocity of the corpuscle electron in the laboratory frame, whereas the corresponding phase speed of the electromagnetic wave of the circulating photon exceeds the ultimate speed of light. The corpuscle electron holds its own Minkowski space-time. That is why the external electric and magnetic fields can interact with the electron and change its kinetic paths in the laboratory frame while simultaneously modifying the phases of the oscillation in different modes in internal frame of the corpuscle electron by the Doppler effect. In the direction of the moving velocity of the corpuscle electron, a blue-shift of the frequency of the rotating vorticity photon would occur, and a red-shift of its frequency would also occur in the direction perpendicular to the velocity orientation.

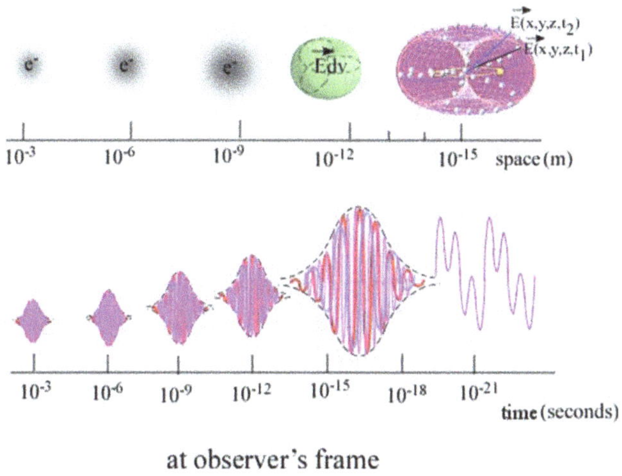

Figure 1.16: The feature of an electron at different length scales and timescales.

The spatial length scale and time duration are significant parameters for the inert frames with relativistic motion because of the Lorenz transformation or Doppler phenomenon. Figure 1.16 shows the features of an electron at different length scales and timescales in the laboratory frame.

1.4.2 Wave function and dynamic parameters of a moving corpuscle electron

As mentioned in the previous section, the electron is a physical entity with a tiny spatial volume as a corpuscle, in which there exists a confined, twisted, circulating photon creating the mass, charge, and spin of the electron. The corpuscle electron has its own spatiotemporal frame, which is different from the laboratory frame. The kinetic parameters of the photon in the moving corpuscle electron with velocity v ($v < c$, where c is the light speed) can be described in the laboratory frame using the Lorentz transformation. The corpuscle electron is usually understood as a material particle in the laboratory frame, but the physical reality is an assembly of tracks of a circulating photon with a spatial periodicity of $\lambda_c \approx 10^{-13} - 10^{-18}$ m and a time period t_0, ranging from 10^{-22} to 10^{-43} per second. The particle and wave duality of an electron can be understood as the relativity phenomenon of the confined electromagnetic energy flux between the laboratory frame and the internal frame of the corpuscle electron. If we see the whole assembly as a physical entity, then the electron is a tiny particle with mass, charge, and spin. However, if we approach spatiotemporal range (e.g., 10^2–10^3 h and 10^{-12} s), in which the internal structure of the corpuscle electron appears to be considered, the characteristics of electrodynamics of the electromagnetic vortices field for a photon must be used. This means that the electron exhibits features of an electromagnetic wave, such as interference and diffraction characteristics. Figure 1.17 clearly demonstrates these features.

Physical parameters of an electromagnetic field are intensity of the electric (E) and magnetic (B) fields, and the energy density of the electromagnetic field in the laboratory frame is

$$U_0 = \frac{1}{8\pi}\int \left(E_0^2 + B_0^2\right) dVol$$

If this electromagnetic field in the corpuscle electron moves with the velocity v and the fields E_0 and B_0 still hold in the laboratory frame, the nonrelativistic limit of Lorentz transformation of the electromagnetic field would be given as

$$E = E_0 - \frac{v}{c} \times B_0 \quad B = B_0 + \frac{v}{c} \times E_0$$

The field momentum related to this moving corpuscle electron would be

$$P_f = \frac{1}{4\pi c}\int E \times B dVol$$

This field momentum orientation changes with time while the photon's circulation creates mass, charge, and spin of the electron. The field parameters do not explain the kinetic and dynamic parameters of the circulating photon. Therefore, we must use another function that contains the kinetic parameters of the dynamic energy of the rotating vorticity photon to express the corpuscle electron system. As mentioned earlier, the kinetic process of the circulating photon has two basic circular motions: (1) circulating photon with energy $\mathcal{E} = mc^2$ and (2) rotating around a circle located in a perpendicular plane as shown in Figure 1.17. Both movements are in two dimensions, $S(2) \times S(2)$. The path of tracks of the circulating photon with elapsed time is a periodic function. The field momentum varies around a composed velocity direction and the assembled moments form a distorted torus.

The relationship between position (q) and momentum (p) or the angular momentum (p_θ, p_φ) with the energy of the circulating photon to express the motion of the electromagnetic field in the corpuscle electron may appear as shown in Figure 1.18.

As M.V. Berry et al. showed, the field momentum and position of the circulating photon on a periodic closed curve path in phase space may be used to describe the dynamic process by defining an angular action function as

At the inert frame of a corpuscle
with velociry V correspoding to observer inert frame

$$\vec{P}(t_0) = (c\vec{E} \times \vec{B})$$

momentum or flux of
electromagnetic energy
Maxwell equation
(only given one dimension)

$$\frac{\partial^2 E}{\partial^2 x} - \frac{1}{c^2}\frac{\partial^2 E}{\partial^2 t} = 0$$

Due to the photon moving
with path on a curved surface,

$$\frac{\delta p}{\delta v} = \delta m$$

$$t = \frac{t_0}{\sqrt{1-\left(\frac{V}{c}\right)^2}}$$

From obsever inert frame
The corpuscle is a physic entity with a finited volume having mass, charge,and spin as an ensemble of electromagnetic energy. This ensemble is a physic energy entity with total energy, $\mathcal{E}^2 = p^2 c^2 + m^2 c^4$.
The description of this physic entity is a wave function ψ, which follow the Klein-golden equation, $\nabla^2 \psi - \frac{m^2 c^2}{\hbar^2}\psi = \frac{1}{c^2}\frac{\partial^2 \psi}{\partial^2 t}$
when the velocity v << c then it follow the Shrödinger equation,

$$-\frac{\hbar^2}{2m}\nabla^2 \phi = i\hbar\frac{\partial \phi}{\partial t}$$

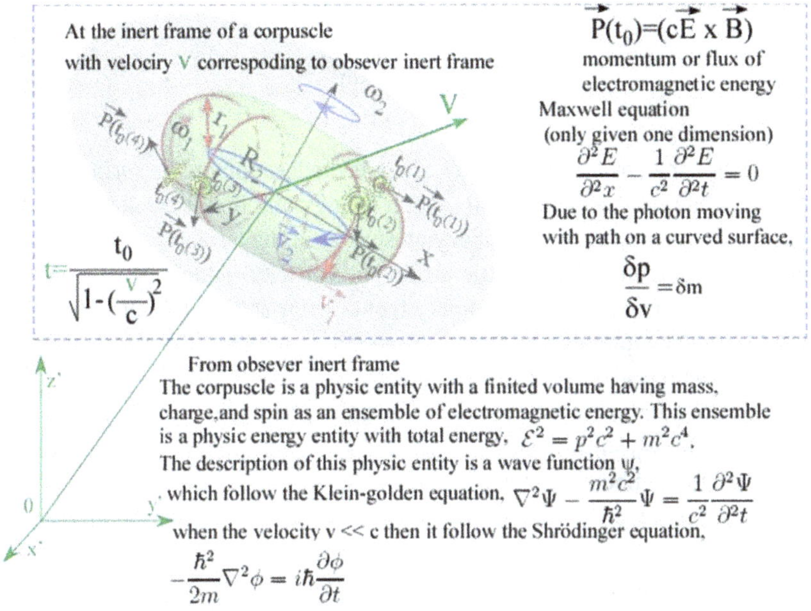

Figure 1.17: The particle and wave duality of an electron and its description.

$\varepsilon = \frac{1}{8\pi}(B^2 + E^2)$
$= \sqrt{(pc)^2 + m_0 c^2}$

$\mathcal{E}(R/r,\theta,\varphi)$

$P(p_x, p_y, p_z)$
$\varepsilon_0 = m_0 c^2$

δq
δp $\delta q \delta p = 2\pi h$

$q(x,y,z)$

(a)   (b)

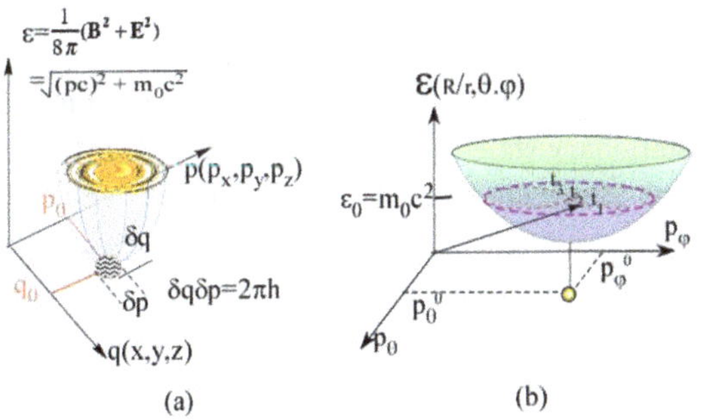

Figure 1.18: The relationship between the energy of the circulating photon and position (q) and momentum (p). (a) Phase space: position (q) and momentum (p), (b) at angular momentum (p_θ, p_φ) space.

$$I_i(q,p) \equiv \frac{1}{2\pi}\oint_{\gamma i} p(q)\cdot dq \quad p(q,E) = \pm\sqrt{\{2m(E - V(q))\}}$$

Track of the circulating photon on the surface of a torus can be expressed as angle variables $\vartheta(q,p)$ defined by

$$\theta = \nabla_I S(q, I)$$

where S(q,I) is the position-dependent action

$$S(q, I) = \int_{q_0}^{q} p(q^{'}, I)\cdot dq^{'}$$

where q_0 is the starting position of the track and q is the location on the surface of the torus at any elapsed time. Tangents of this action at any point on the track path of the circulating photon are related to the velocity on surface of the torus.

For the first circulation of the kinetic process or vorticity, value of the I_1 is equal to $\mathcal{E}/\omega$, but if $\mathcal{E} = nh\omega$, then $I_1 = nh$. This implies that angular action variables are equal to a quantum number multiplied by Planck's constant. Therefore, the angular action variables become temporal evolution phase factors.

For the second motion, the rotation of the vorticity photon, its angle action variables are $I_2 = L_\varphi = mR^2\ w_\varphi = mv_\varphi R = p_\varphi R$, which the kinetic parameter that do not directly correspond to the photon's energy.

From this analysis, we may write a function comprehensively describing the dynamic and kinetic processes of the electromagnetic field energy flux of a vorticity photon in the confined space and elapsed time as follows:

$$\Psi(r,\theta,\varphi,t) = F(r,\theta,\varphi,t)\exp[i/h(pr - (\varepsilon_v + \varepsilon_0)t)]$$
$$= F^{'}(x,y,z,t)\exp\left[i/h\left(p_x x + p_y y + p_z z\right) - (\varepsilon_v + \varepsilon_0)t\right]$$

where F′(x,y,z,t) is F(r,θ,φ,t) multiplied by the Jacobian transformation, and $\mathcal{E}_v$ is the energy gain of the corpuscle electron moving at velocity v in the laboratory frame; $\mathcal{E}_0$ is the vorticity photon's energy in a stationary state in the internal frame of the corpuscle electron. As J.G. Williamson et al. indicated, an increase or decrease in $\mathcal{E}_0$ will lead to the absorption or radiation of electromagnetic field energy with a frequency as high as 10^{22} Hz, but the kinetic movement at velocity v does not change $\mathcal{E}_0$. It modifies the frequency of the circulating photon except for strong interaction with the atom's nucleus; $\mathcal{E}_v$ is the kinetic energy gained from the circulating photon due to the Doppler effect. Therefore, the energy $\mathcal{E}_v$ represents the kinetic energy of the whole corpuscle electron and can be called "de Broglie wave energy." Based on this argument, we can write down as

$$\begin{aligned}\Psi(r,\theta,\varphi,t) &= F(r,\theta,\varphi,t)\exp[i/h(pr-\varepsilon_v t)]\exp(i/h(\varepsilon_0 t))\\ &= F'(x,y,z,t)\exp\left[i/h\left(p_x x+p_y y+p_z z\right)-\varepsilon_v t\right)]\exp(i/h(\varepsilon_0 t))\\ &= \Psi_v(r,\theta,\varphi,t)\exp(i/h(\varepsilon_0 t))\end{aligned}$$

The wave function $\Psi_v(r,\theta,\varphi,t)$ is the same as before, when we got one $\Psi_v = \Psi_0 \exp [i/\hbar (pr-\mathcal{E}_v t)]$, and it clearly means that (1) the wave function of the corpuscle electron describes the kinetic and dynamic characteristics of a physical entity, which consists of the assembly of the twisted, circulating tracks of the photon on the topologic surface of a torus. (2) The position of the physical entity as a corpuscle (as de Broglie used it) in the laboratory frame may be described as a Dirac delta function or a Green's function, and the extension of the Dirac delta function may be expressed as $10^{-12} \sim 10^{-5}$ m. (3) The phase exp(i/h(pr)) expresses the rotational movement of the twisted, circulating photon in the laboratory frame. (4) The phase exp (i/h (ε_vt)) reveals the distribution of the additional energy induced by the Doppler effect of the circulating photon at an elapsed point in the internal frame of a corpuscle electron.

It is worth emphasizing that the electromagnetic field of the vorticity photon follows Maxwell law, and this field's interaction with the external field should obey laws of electrodynamics mechanics. Because of the circulating motion of the photon on the curved surface of a torus, a mass should be created from the dynamic curving motion of the electromagnetic field flux (vorticity electromagnetic field), and assemblies of the electromagnetic field energy flux as a corpuscle with mass will obey the Klein-Gordon equation or the Schrödinger equation. Therefore, the wave function follows the Schrödinger equation.

The basic property of the Dirac delta function is

$$\int_{\infty}^{\infty} \delta(x)f(x)dx = f(0)$$

If the function f(x,y,z) is the trajectory of the corpuscle or torus, at any position

$$F(x_0,y_0,z_0,t) = \int \delta(x-x_0, y-y_0, z-z_0,)F(x,y,z,t)dxdydz$$

It is obvious that $\Psi_v \bullet \Psi_v{}^* = |F(x_0,y_0,z_0,t)|^2$. This tells us that normalization of the wave function contains the hidden meaning of the internal structure of the corpuscle electron.

It is useful to indicate that the corpuscle electron may be imaged as a wave packet in which many plane electromagnetic waves with different field momenta combine to form a wave packet. Distributions of the field momenta usually have a Gaussian function, which can approach limits as a Delta function.

1.4.3 Wave function of a free corpuscle electron

Electron microscope is an instrument used to explore the micro-world by using free moving electrons, which are steered by electric and/or magnetic lenses and the sample. Therefore, if you want to steer these free moving electrons, you have to know the concrete spatial positions and transient times that are physical kinetic parameters of the real entity as a corpuscle electron. As mentioned in the previous sections, an electron is not a plane wave or a material particle, but a quantum system in S(2) × S(2) topological space in which electromagnetic energy flux periodically circulates around two perpendicular circles, as shown in Figure 1.19. The transient electric field intensity, $|E| = E_0\cos[\omega_{db}\,(xv_t/c^2\text{-}t)]$, is a physical parameter of the corpuscle electron, but its average over time in this internal space would give the electric field of an electron in the laboratory frame. If we consider the electron to be at rest, that means $v_t = 0$, then $|E| = E_0\cos\,[\omega_{db}(t)]$ or $|E|^2 = E^2_0\cos^2\,[\omega_0(t)]$ due to $\omega_{db}\,(t) = \omega_0(t)$ at rest. This gives the electromagnetic energy of the photon, as $\mathcal{E} = h\omega_0$. Based on this analysis, it may be clear that: (1) steering the velocity of an electron should imply modifying the velocity of the circulating photon in the internal frame of a corpuscle electron or, in other words, steering the periodicity frequency of the circulating photon, which means modifying wavelength of the de Broglie matter wave.

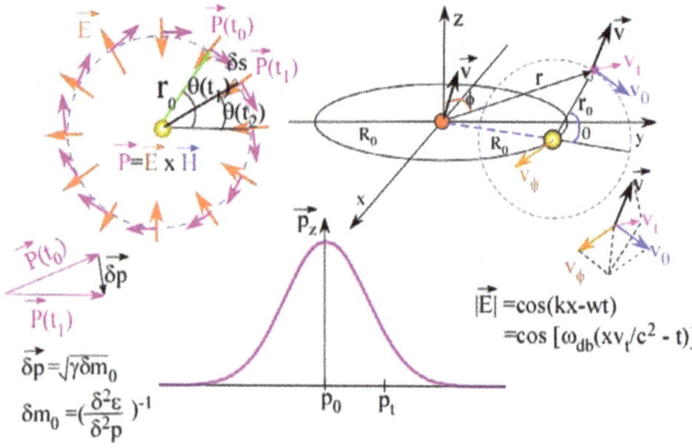

Figure 1.19: Illustrating the mass created by the curved motion of the electromagnetic field momentum and the distribution of the momentum projected on the z-axis. The kinetic velocity of a photon's track at an elapsed time is also shown. The electric field is a periodic function of the velocity of a moving electron at a transient time.

(2) The intrinsic energy of an electron is the electromagnetic field energy of the circulating photon at rest in a corpuscle electron. This dynamic motion system may be steered by an external field or the collision of two systems. This dynamic motion system of the photon is stable except in the case of a collision with extra-high energy.

(3) The Compton wavelength is related to the internal system of a corpuscle electron, but the de Broglie matter wave is related to the velocity of the corpuscle electron as it relates to the dynamic rotating vorticity photon. Therefore, the matter wave or de Broglie wave is a relativistic phenomenon of the dynamic circulating electromagnetic field flux between the laboratory frame and the internal frame of a corpuscle electron.

(4) Because the dynamic circulating photon is confined in a space formed by two perpendicular, two-dimensional motions, S(2)xS(2), and the moving system with velocity, v_t, the observer in the laboratory frame may see a particle existing in a tiny spatial volume. The tracks' trajectories of the circulating photon in the laboratory frame may be described as a Dirac delta function, which is the solution of the dynamic motion equation. Such as

$$\frac{1}{2\pi}\int_{-\infty}^{+\infty} e^{-ik(y-x)}dk = \delta(y-x)$$

$$f(x) = \int_{-\infty}^{+\infty} \delta(x-y)f(y)dy$$

$$\delta(x-x') = \frac{1}{2L}\sum_{n=-\infty}^{\infty} e^{in\pi(x'-x)/L}$$

(5) If systems of the corpuscle electron have total energy as

$$\varepsilon_{tot} = mv_t^2 + m_0c^2\left(1-(v_t/c)^2\right)^{1/2}$$

phase velocity of the rotating vorticity photon,

$$v_{ph} = \omega/k = h\omega/hk = \varepsilon_{tot}/p = \left[mv_t^2 + m_0c^2\left(1-(v_t/c)^2\right)^{1/2}\right]/mv_t = c^2/v_t$$

Then the relationship, $v_t \bullet v_{ph} = c^2$, is de Broglie's duality of particle and wave motion, in which v_t is the particle's velocity or group velocity, and v_{ph} is the phase velocity of a wave. It is clear that if the particle's velocity, v_t, is changed in magnitude or orientation, then the phase velocity, v_{ph}, must change simultaneously, and vice versa.

At an elapsed time, if the track's location of the circulating photon at a torus surface is related to a phase ($\varphi = \omega t'$) of the electromagnetic harmonic oscillator, then this phase propagating velocity, i.e., the phase velocity $v_{ph} = c^2/ v_t$, and if the angular frequency, ω, of the electromagnetic energy flux of the rotating vorticity photon at the internal frame is modified, then the phase velocity is modified. Modifying the phase velocity would change the group velocity of the corpuscle electron. This internal kinetic motion can be described as

$$\exp\left[i/h\left(p_xx+p_yy+p_zz\right)-\varepsilon_vt\right)\right]\exp((i/h)\varepsilon_0t) = \exp[i\omega_{db}(rv_t/c^2-t)]\exp[(i/h)\varepsilon_0t]$$

It is clear that position r, velocity v_t of the corpuscle system, and the elapsed time t are physically real parameters being used when steering the corpuscle electron.

In summary, wave function of an electron, such as

$$\Psi(r, \theta, \varphi, t) = F(x(t), y(t))\delta(z - v_t t)\exp\left[i/h\left(p_x x + p_y y + p_z z\right) - (\varepsilon_v t)\right]$$

describes an electron as an isolated corpuscle, in which the confined dynamic circulating photon is nested in a spatiotemporal Minkowski frame. The confined dynamic circulating photon has $\mathcal{E} = h\omega = (p^2c^2 + \gamma m_0{}^2c^4)^{1/2}$, in which the mass emanates from the vorticity of the electromagnetic field momentum rotates around mass center of the corpuscle, and the corpuscle flies at velocity v_t in the laboratory frame. This combined kinetic motion of the linear and curving (vorticity) dynamic oscillation of the electromagnetic energy flux is the physical character of an electron in electron microscopy.

References

Aspden, R.S., Padgett, M.J. and Gabriel, C. "Spalding Video recording true single-photon double-slit interference." American Journal of Physics. 84, 671, (2016). doi: 10.1119/1.4955173,.

Bach, R. and Pope, D. "Liou Sy and Batelaan H Controlled double-slit electron diffraction." New Journal of Physics. 15, 033018, (2013).

Barwick, B., Gronniger, G., Lu, Y., Liou, S.Y. and Batelaan, H.A. "measurement of electron–wall interaction using transmission diffraction from nanofabricated gratings." Journal of Applied Physics. 100, 074322, (2006).

Bass, P.G., "Derivation of the Schrodinger, Klein-Gordon and Dirac equations of particle physics via classic method" www.relativitydomains.com (January 2016).

Berry, M.V. and Pragya, S. "Hamiltonian curl forces." Proceedings of the Royal Society A: Mathematical, Physical and Engineering Sciences. 471, 0002, (2015).

Bourgoin, R.C. "On the Formation of photon Toroids." Advanced Studies in Theoretical Physics . 2(5), 215–227, (2008).

Caesar, C. "Model for understanding the substructure of the electron." Nature Physics. 13(7), 1, (2009).

Daywitt, W.C. "The Compton radius, the de Broglie radius, the Planck Constant, and the Bohr orbits." Progress in Physics. 2, 32, (2011).

Louis., D.B. New Concept of Light. Translated by Delphenich, D.H., Editors 6 Rue de la Sorbonne, 6: Paris Hermann & Co., (1934).

Louis., D.B. The Theory of Particles of Spin ½ (Dirac Electrons). Translated by Delphenich, D.H., Quai des Grands-Augustins: Paris Gauthier-villars, printer-editor library of the bureau of longitudes of L'école polytechnique p. 55, (1952).

de Broglie, L.-V. On the Theory of Quanta. PARIS, A translation of RECHERCHES SUR LA TH´EORIE DES QUANTA (Ann. de Phys., 10e s´erie, t. III (Janvier-F ´evrier 1925).by: A. F. Kracklauerc, AFK, pp. 1892–1987, (2004).

Frabboni, S., Frigeri, C., Gazzadi, G.C. and Giulio, P. "Two and three slit electron interference and diffraction experiments." American Journal of Physics. 79, 615, (2011). doi: 10.1119/1.3560429.

Frabboni, S., Gabrielli, A., Gazzadi, G.C., Giorgi, F., Matteucci, G., Pozzi, G., Cesari, N.S., Villa, M. and Zoccoli., A. "The Young–Feynman two-slits experiment with single electrons: Build-up of the

interference pattern and arrival-time distribution using a fast-readout pixel detector." Ultramicroscopy. 116, 73–76, (2012).
Frabboni, S., Gazzai, G.C. and Pozzi, G. "Young's double-slit interference experiment with electrons." American Journal of Physics. 75, 1053–1055, (2007).
Frabboni, S., Gazzai, G.C. and Pozzi, G. "Nanofabrication and the realization of Feynman's two-slit experiment." Applied Physics Letters. 93, 073108, (2008).
Frabboni, S., Carlo Gazzadi, G., Vincenzo, G. and Giulio, P. "Elastic and inelastic electrons In the double-slit experiment: A variant of Feynman's which-way set-up." Ultramicroscopy. 154, 49–56, (2015).
Gauthier, R. "The electron is a charged photon with the de Broglie wavelength." www.superluminalquantum.org.
Michel, G. and Gondran, A. "Measurement in the de Broglie-Bohm interpretation: Double-slit, Stern-Gerlach, and EPR-B." Physics Research International. Article ID 605908, 16, (2014). http://dx.doi.org/10.1155/2014/605908.
Grössing, G., Fussy, S., Johannes, M.P. and Schwabl, H. Relational Causality and Classical Probability: Grounding Quantum Phenomenology in a Superclassical Theory. Austrian Institute for Nonlinear Studies Vienna, (3–6 Oct 2013).
Kanarev, P.M. "Plank's constant and the Model of the electron." Journal of Theoretics.
Kocsis, S., Boris, B., Ravets, S., Stevens, M.J., Mirin, R.P., Krister Shalm, L. and Steinberg, A.M. "Observing the average trajectories of single photonsin a two-slit interferometer." Science. 332, 1701, 3 JUNE, (2011).
Marquet, P. "On the physical nature of the de Broglie Wave." Progress in Physics. 12, 318, (2016).
Merli, P.G., Missiroli, G.F. and Pozzi, G. "On the statistical aspect of electron interference phenomena Am." Journal of Physics. 44, 306–307, (1976).
Namsrai, K. Torus Theory, (2001), http://www.ictp.trieste.it/~pub-off
Oriols, X. and Mompart, J. "Overview of bohmian mechanic." In: Applied Bohmian Mechanics: From Nanoscale Systems to Cosmology, editorial Pan Stanford Publishing Pte.Ltd, pp. 15–147, (2012).
Sanz, A.,.S. and Miret-Art'es, S. "A trajectory-based understanding of quantum interference." Journal of Physics A: Mathematical and Theoretical. 41, 435303, (2008).
Shanahan, D. "The de Broglie Wave as Evidence of a Deeper Wave Structure." https://www.researchgate.net/publication/280713502.
Tonomura, A., Endo, J., Matsuda, T., Kawasaki, T. and Ezawa, H. "Demonstration of single-electron buildup of an interference pattern Am." Journal of Physics. 57, 117–120, (1989).
Williamson, J.G. and Van der Mark, M.B. "Is the electron a photon with toroidal topology?." Annales de La Fontation Louis de Broglie. 22(2), 133, (1997).
Williamson, J.G. A New Theory of Light and Matter. Marseille, France: FFP14, (2014).
Williamson, J.G. and Leary, S.J. Absolute Relativity in Classical Electromagnetism: The Quantization of Light. San Diego: SPIE optics+photonics, pp. 9570–41, (9–3 Aug 2015).
Williamson, J.G. "On the nature of the photon and the electron," companion paper to this one for August (2015).
Zürcher, U. "What is the frequency of an electron wave?." European Journal of Physics. 37, 045401, (2016).

Chapter 2
Intrinsic nature of a corpuscle electron in an electron microscope

2.1 Electron in the electron gun of an electron microscope

As is well known, the electrons escaping from the filament of an electron gun, for example, the tungsten V-shaped filament or tip of a lanthanum hexaboride (LaB_6) or gallium arsenide phosphide (GaAsP) single crystal or carbon nanotube, have a nearly stationary state or zero velocity. The distribution of the escaped electrons around zero velocity may control the monochromatic property of electrons, which gives better coherency.

For an electron, as a corpuscle of confined electromagnetic energy, its total energy is

$$\begin{aligned}\varepsilon_{tot} &= m_0 v^2/\left(1 - v^2/c^2\right)^{1/2} + m_0 c^2\left(1 - v^2/c^2\right)^{1/2} \\ &= mv^2 + m_0 c^2\left(1 - v^2/c^2\right)^{1/2}\end{aligned}$$

This relation is called the "Planck-Laue formula," which unveils the relationship between kinetic energy (as mv^2) and dynamic energy (as $m_0c^2(1\text{-}v^2/c^2)^{1/2}$) of the corpuscle electron that plays a key role in modern physics.

If the kinetic energy of a moving corpuscle electron is ε_v, then

$$\varepsilon_v = mv^2 = v^2/c^2\left(mc^2\right) = v^2/c^2\varepsilon_{tot}$$

This means the kinetic energy of a corpuscle electron is only part of the energy of a corpuscle electron.

If the velocity of the corpuscle electron, v_t, is the velocity in the laboratory frame and is less than the speed of light, c (i.e., $v < c$), the kinetic energy would be

$$\varepsilon_k = \varepsilon_{tot} - m_0 c^2 = mv^2 - m_0 c^2\left[1 - \left(1 - v^2/c^2\right)^{1/2}\right]$$

If the velocity of the corpuscle electron is $v \ll c$, the roots can be expanded to obtain

$$\varepsilon_{tot} = \varepsilon_k + m_0 c^2 = 1/2 m_0 {v_t}^2 + m_0 c^2$$

It tells us that the energy of a corpuscle electron's motion only constitutes part of its total energy, and the internal energy is the fundamental energy of the corpuscle electron; the electromagnetic field energy is inherent or intrinsic. It is important to remember that the motion of a corpuscle electron is in the laboratory frame, but the internal circulating electromagnetic field is dynamic motion within the internal

https://doi.org/10.1515/9783111449333-002

frame of the corpuscle electron, and the electron's energy carrier is the circulating photon (or the vorticity photon). Therefore, if we measure the characteristic parameters of the circulating electromagnetic energy, we have to consider the relativistic transformation between the two inertial frames.

Energy core of an electron is $\varepsilon_0 = m_0c^2$ or $\varepsilon_0 = h\nu_0$ that is the energy of the electromagnetic field of a photon, but for an observer in the laboratory frame, the frequency of the electromagnetic field is $\omega = \omega_0\left(1 - v^2/c^2\right)^{-1/2}(\omega = 2\pi\nu)$ that is related to the kinetic motion of the corpuscle electron, as shown in Figure 2.1.

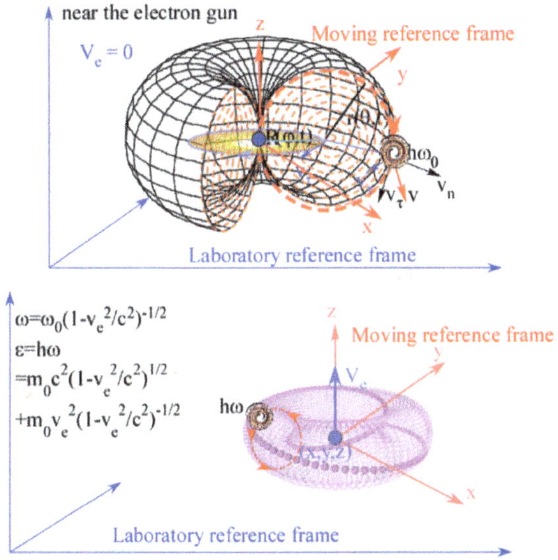

Figure 2.1: At rest, the corpuscle has two motions: (a) the twisted circulation of a photon with $\mathcal{E} = \hbar\omega_0$ and (b) the rotation of the whirling photon with phase velocity $\mathbf{v}_{ph}$, which is nearly infinite. In motion, a corpuscle's energy and angular frequency have to relate to the velocity, v_e, of the corpuscle, and the morphology of the corpuscle is modified by the difference in phase velocity in its internal frame.

The de Broglie wave length λ usually used in electron microscopes is based on the kinetic energy of a moving corpuscle electron in the laboratory frame. The phase of dynamic periodic motion of the electromagnetic energy flux in the internal frame of the corpuscle electron is also influenced by the velocity of the corpuscle electron. If we accept this view, the phase between the electromagnetic field and the motion of the corpuscle electron should be the same if there is no external interaction, for example, electric and magnetic fields. It should be equal in the laboratory and internal frames of the moving corpuscle. The rotation phase can be expressed as

$$\varphi = \nu_0 t = 2\pi\left(m_0c^2/h\right)t = 2\pi m_0c^2\left(t - v_t x/c^2\right)/h\left(1 - {v_t}^2/c^2\right)^{1/2} = \nu t - kx$$

where $k = 1/\lambda = 2\pi m_0 v/h(1 - {v_t}^2/c^2)^{1/2}$ and $\nu = \nu_0(1 - (v_t/c)^2)^{-1/2}(\omega = 2\pi\nu)$.

This formula connects a particle's (corpuscle's) momentum with the phase of its rotating electromagnetic wave of the vorticity photon. The wave vector is related to the vector of velocity of the corpuscle electron. In other words, wave vectors k of electromagnetic waves in the electron's internal frame are a function of the velocity of a corpuscle electron.

Corpuscle electrons have a nearly spherical morphology, and their momentum has a minimal value of dynamic electromagnetic energy flux $p_0| = m_0c = h\omega_0/c$. The electric field of the circulating photon can be expressed as

$$E_r(\theta, t) = E_r{}^0\exp[i(\omega_0 t - (r\omega_0/c)\theta)] = E_r{}^0\exp[i(\omega_0 t - k_{(r\theta)}\theta)]$$

where $k_{(r\theta)} = r\omega_0/c$ is the wave vector, and $E_r{}^0$ is the circulating electric field vector with angular velocity ω_0 and tangential linear velocity $v_\theta = \omega_0 r/k_{(r\theta)} = c$ and $c = (\mu_0\varepsilon_0)^{1/2}$. It shows that the circulating photon is circulating at the speed of light c, as mentioned in Chapter 1. The enveloped spatial volume of the circulating photon exhibits the electric charge, which can be calculated by the Maxwell equation:

$$\int_{\text{surface}} E_r(\theta, t)d\theta dt = \rho_e/\varepsilon_0$$

where ρ_e is the electricity of the corpuscle electron, which emanates from its enveloped surface, and its electric field is $E_r{}^0 = \rho_e/4\pi\varepsilon_0 R^2$ at rest in the laboratory frame. Interaction between the corpuscle electron and its external fields (e.g., in an electron gun or magnetic lens) exerts an electric field $E_r{}^0$ on the corpuscle electron, changing its velocity (orientation and value of the momentum). However, the velocity of the corpuscle electron is the group velocity of the internal dynamic circulating wave of the photon. The velocity change of the corpuscle electron complies with the de Broglie relation, $v_g \bullet v_{ph} = c^2$, which implies that the phase velocities of the circulating photon have been modified in the internal frame of the corpuscle electron.

In an electron microscope, electrons escaping from the cathode should have different locations (depending on the geometry of the tip) and temporal-energy ranges (related to the excited process, such as thermal and electric field extraction, or laser radiation). The escaping electrons might have tiny energy differences with different escape velocities. That is the starting state of the electron momentum as $d\varepsilon/dv_i = p_i$. Momentum of these electrons is $p_i = m_0c = (p^i{}_x{}^2 + p^i{}_y{}^2 + p^i{}_z{}^2)^{1/2}$. Because they are nearly stationary, the components of different directions are equal to $p^i{}_x = p^i{}_y = p^i{}_z$. The starting wave function may be as $\psi = \psi_0\exp(i/h(p_i r))$ and $\psi_0 = E_r{}^0\exp[i(\omega_0 t - k_{(r\theta)}\theta)]$, where phase $\varphi_i = i/h(p_i \bullet r)$ is the phase of rotating motion in the internal frame of the corpuscle electron, but the phase $\varphi = [i(\omega_0 t - k_{(r\theta)}\theta)]$ is the phase of the circulating electromagnetic field of the photon. However, a corpuscle electron in an accelerating static electric field is immediately accelerated, and in the shortest temporal range, $\delta t = (d/eV)(\delta\varepsilon/v_i)$ (where d is the distance between cathode and anode, e is the charge of the electron, V is the acceleration voltage, v_i is the initial velocity, and $d\mathcal{E}$ is the constant energy spread), the velocity, v_t, of an accelerated electron may approach about light speed (e.g., near 0.69 light speed for 200kV, which is equal to 2×10^8 m/s). The acceleration time may be $\delta t = 1.8$ ps for 100 kV and 0.6 ps for 300 kV. The corpuscle electron after the gun flies into free space, and the flying time between the gun and specimen for the electron with different escape energies is about a few ns (nanoseconds). The electron's velocity is not influenced by the flight time, so the corpuscle electron is flying sequentially.

As discussed in Chapter 1, a corpuscle electron can be described as a function

$$\Psi(r,\theta,\varphi,t) = F(r,\theta,\varphi,t)\exp[i/h(pr-(\varepsilon_v+\varepsilon_0)t)] = \Psi_0\exp[i/h(pr-\varepsilon_v t)]$$
$$= E_r{}^0\exp[i(r\omega_0/c)\theta-(\omega_0 t)] = E_r{}^0\exp\left[i(k_{(r_\theta)}\theta-\omega_0 t)\right]$$

After flying out of the gun, the corpuscle electron's energy is conserved, but the kinetic motion may be modified by, for example, a magnetic lens or the sample.

The kinetic movement will form a trajectory based on the action or eikonal function S at each location of tracks on the trajectory. The system of kinetic conservation obeys the Maupertuis law: $\nabla S + e\Lambda = mv_t$, where v_t is the transient velocity at a point on a trajectory. In a magnetic lens, Λ is stable, and the lens can change direction but not $|v_t|$, yet the total energy of the corpuscle electron is still conserved. However, in a magnetic lens, the atomic core potential in a sample can also modify the transient velocity of the corpuscle electron at a location on the trajectory. In free space, though, the transient velocity is always perpendicular to the surface of the action or eikonal function, which is the phase factor $[(pr - \varepsilon_v t)/h]$, where the momentum, **p**, and position, r, are the values measured in the laboratory frame (the electron microscope); $\omega_v = \varepsilon_v/h$ is the de Broglie wave's angular frequency and is an additional frequency of the rotation and circulating photon in the internal frame of the corpuscle electron due to the electron's motion. This motion speed is called the "group velocity" and is related to the phase velocity of the rotating and circulating electromagnetic wave of a photon in the internal frame of a corpuscle electron. The phase of the rotating motion should be the same in any inertial frame based on the principle of relativity, but different inertial frames have their own coordinate systems with different relative velocities. If coordinates x, y, z, t are in the laboratory frame and x′, y′, z′, t′ are in the internal frame of the corpuscle, the corpuscle electron is flying along the optic axis of the microscope as the z-axis. Then we have

$$z' = \gamma(z - v_t t),\ y' = y,\ x' = x \text{ and } t' = \gamma[t - (\beta/c)z)]$$

If the electric field at rest is

$$E_r^0\exp[i(r\omega_0/c)\theta-(\omega_0 t)]$$

then when the corpuscle electron moves at speed v_t in z direction in the laboratory frame, the electric field should be as

$$E_r^0 \exp[i\omega'_0/c(\beta z + r'\theta) + \omega'_0 t]$$

where $\omega'_0 = \gamma\,\omega_0$, $r' = r/\gamma$, and $\beta = v_t/c$. It indicates that the rotation frequency is increasing (in this case along the z-axis), and the electromagnetic wave length is decreasing correspondingly, and the z′-axis of internal coordination of the corpuscle electron has contracted proportionally. However, only the z-z′ coordination of the laboratory frame has contracted. Because $\omega'_0 = \gamma\omega_0$, energy of the corpuscle electron is increased $h\omega'_0 = \gamma h\omega_0$ and the additional increased energy is its kinetic energy. Simultaneously, the mass is increased, and the wave length is decreased in the laboratory frame. However, based on phase conservation at two inertial frames and $\varphi = [i\omega'_0/c(\beta z + r'\theta) + \omega'_0 t] = \text{constant}$, ω'_0/c is the wave vector k'. The $(\beta z + r'\theta)$ factor has complex coordination characters, such as βz, which is the coordinate in the laboratory frame, but the $r'\theta$ factor is the coordinate in the internal frame of the corpuscle electron.

Because the electron is moving but the gun, lens, and sample are stationary, the observer in the laboratory frame can only see the (βx) part of the coordination and the de Broglie wave, which is the additional oscillation of the circulating photon in the internal frame of the electron, hidden in the phase $r'\theta$. If we differentiate both sides of the phase, we can obtain $\beta v_z + r'\omega_\theta = c$, where ω_θ is the angular velocity in the internal frame of a corpuscle electron, and v_z is the moving velocity in the laboratory frame. Using $r_0\omega_\theta = v_\theta$ and $\beta v_x + v_\theta/\gamma = c$, we can have $(v_z{}^2 + v_\theta{}^2)^{1/2} = c$ and $v_\theta = c(1-\beta^2)^{1/2}$ (note that $\beta = v_t/c$). It indicates that the flying velocity increased, and the internal rotation velocity decreased accordingly. That is why if the linear velocity of a corpuscle electron approaches the speed of light, it results in a linearly propagated photon with light speed. When the electron is at rest, the tangent velocity of the circulating photon is at the speed of light.

The velocity, v_t, of a corpuscle electron is a measurable kinetic parameter, but v_θ is the tangent velocity of a rotating circulating photon at the internal frame of a corpuscle electron moving at velocity v_t, and not measurable in the laboratory frame. However, these two velocities are connected to each other by the phase factor $\varphi = [ik(\beta z + r'\theta) + \omega'_0 t]$. Its differentiation with respect to time should be zero because it is constant for any inertial frame as $d\varphi/dt = 0$. Therefore, we can have $\omega = v_{ph} \bullet k$ and $\omega/k = \lambda v = v_{ph}$. This is a fundamental wave-character relationship. It is easy to prove this $v_t \bullet v_{ph} = c^2$, which is de Broglie's particle-wave duality of the matter wave. This equation multiplies two vectors to get the scalar value, which means that the measurable direction modification of the velocity of the corpuscle electron may emanate from rectifying the corresponding phase velocity variation, which emanates from the angular frequency modulation or the wave vector variation of the electromagnetic wave field of the photon.

2.2 Electron flying along the optic axis of a transmission electron microscope

Acceleration in a certain direction results in the gaining speed of the corpuscle electron. The gaining speed is not uniform but mainly along a specific direction (e.g., the z direction). Therefore, the momentums of the electron have preferential rather than uniform directions. The ratio between the preferential direction and others is $\alpha = p_x/p_z = p_y/p_z$ or $\alpha = p_r/p_z \left(p_r^2 = p_x^2 + p_y^2\right)$, which dominates the flying direction and trajectory of the electron. This ratio also appears in the difference of de Broglie wave length along different directions, resulting in shape and phase velocity variation of the corpuscle electron, which is manipulated by the interaction between the electron and the external field.

It is worth mentioning that $\alpha = p_r/p_z = hp_r/hp_z = \lambda_z/\lambda_r$ indicates the flying divergent angle and de Broglie wave length of the electrons in the velocity direction, i.e., longitudinal or optic axis, are much shorter than those in the transverse direction. If the divergent angle is around 0.01 rad, then the λ_r may equal to about 100 λ_z. For example, if λ_z is 0.27 nm (i.e., 200kV), then λ_r is about 27 nm.

Figure 2.2 shows the modification of the corpuscle electron's morphology at the electron gun of an electron microscope. As mentioned before, the corpuscle electron is an anti-bunched fermion in the laboratory frame. No two corpuscle electrons can have the same position at the same time. A free-flying electron is in sequential motion, and its energy is conserved along its trajectory if there is no external action (according to Maupertuis's principle). The law of energy conservation allows trajectories to be curved. The path of the corpuscle electron follows the variation principle of Maupertuis for constant total energy. Based on the Lagrange variation principle,

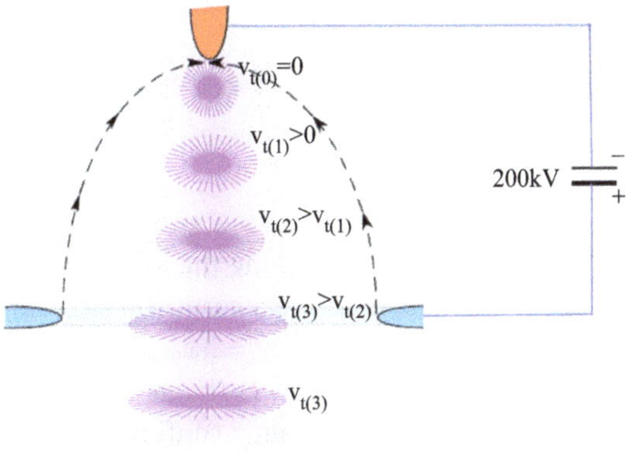

Figure 2.2: The morphology of a corpuscle electron is distorted by a static electric accelerating field, increasing the velocity because the longitudinal velocity is higher than that of the transverse direction.

$$\delta \int L dt = \delta \int [pq - H] dt = 0$$

If $H = \mathcal{E}$, the variation principle of Maupertuis for constant total energy is

$$\delta_{H=\varepsilon} \int [pq - H] dt = \delta_{H=\varepsilon} \int p dq = \delta \int [p(r)|dr/dz| + e\Lambda(dr/dz)] dz = 0$$

where the momentum is $p(r) = m\gamma v$(**r** and **v** are vectors). The Lagrange variation principle applied to a physical trajectory is now as $\delta \int L dz = 0$, and is called an "eikonal." It is usually written as $\exp(\int_0^z L dz)$. If the initial coordinate q_i and the final coordinate q_f have been chosen, the eikonal can be expressed as

$$S(q_i, q_f) = \int p \, dq$$

where integral is from q_i to q_f.

The principle of Maupertuis would be $\delta S = \delta \int_{r(i)}^{r(f)} p(r) dr = 0$ in free space. The direction of the electron trajectories is perpendicular to the surfaces of the constant eikonal, and the momentum of the corpuscle electron is also perpendicular to this surface. In this situation, the de Broglie wave vector $k_{r\theta}$ is also perpendicular to this surface and may be viewed as a vector on the wave front of the de Broglie wave. Electrons flying out of the anode of the gun may have diverged slightly (e.g., 0.01 rad). Therefore, the de Broglie wave vector also has tiny divergences. The flying electron has a shorter longitudinal and longer transversal wave length, which may be why the electron wave is usually viewed as a plane wave. In other words, the surface of the constant eikonal is almost flat.

Corpuscle electron can be described as a wave function:

$$\begin{aligned}\Psi(r, \theta, \varphi, t) &= F(r, \theta, \varphi, t) \exp[i/h(pr - (\varepsilon_v + \varepsilon_0)t)] \\ &= F(r, \theta, \varphi, t) \exp[i/h(\varepsilon_0)t)] \exp[i/h(pr - \varepsilon_v t)] \\ &= \Psi_0(r, \theta, \varphi, t) \exp[i/h(pr - \varepsilon_v t)]\end{aligned}$$

where Ψ_0 $(r,\theta,\varphi,t) = F$ $(r,\theta,\varphi,t)\exp[i/h$ $(\mathcal{E}_0)t)]$ is a function describing the kernel of the corpuscle electron, which contains the space-time dimension of a corpuscle electron, in which the circulating photon's rotating motion holds energy $\varepsilon_0 = h\omega_0 = m_0c^2$. This kernel is an essential part of the confined electromagnetic energy. The motion of the corpuscle electron gives additional kinetic energy to the dynamic confined electromagnetic field. The intrinsic circulating frequency is very high, about $\approx 10^{22}$ Hz. The kinetic energy is equal to $(v^2/c^2)\mathcal{E}_t$, which is related to the velocity of the corpuscle electron in the laboratory frame.

In general, $\Psi_{db} = \Psi_0(r, p, t) \exp[i/h (pr - \mathcal{E}_v t)]$ expresses the wave function of a corpuscle electron. But in this function, the phase factor, $[i/h (pr - \mathcal{E}_v t)]$, (as we mentioned previously) contains the measurable momentum, p, position coordinates, r, and the relative kinetic energy, $\mathcal{E}_v$, and time, t, in the laboratory frame. All of them comprise the corpuscle electron's physical parameters.

These parameters should carry the information codes of the kinetic and dynamic circulating electromagnetic field flux of a photon. But the kernel of the corpuscle electron carries the total energy of the rotating electromagnetic vorticity flux of the photon, which is a real measurable parameter for electron microscopy. The function Ψ_0 (r, p, t) should hold the energy of a corpuscle electron. Because of the nature of electromagnetic field energy, the time-space average energy density of this corpuscle system can be expressed as

$$<\varepsilon_k>_{t,s} = 1/2\mu_0 H_o^2 = 1/2\epsilon_0 E_o^2 = \Sigma_n \varepsilon_{n(k)}$$

and time-space average Poynting vector

$$<S_k>_{t,s} = 1/2\mu_0 H_o^2(\omega/k)z_o = 1/2\epsilon_0 E_o^2(\omega/k)z_o = \Sigma_n S_{n(k)} \cdot \varepsilon_{n(k)}$$

where $\mathcal{E}_{n(k)}$ and $S_{n(k)}$ are the energy and Poynting of nth elapsed component waves of the rotating electromagnetic field of the vorticity photon. E_o and H_o are the time-space average electric and magnetic field intensities of the corpuscle electron. In the laboratory frame, the electric field of an electron is a stationary field (E_0), which is the temporal average of the ultrafast varying electric field E(t) in the time domain (-τ,τ).

Wave function, Ψ, discussed previously, may properly be presented as the electric field E(t) having the following form:

$$E(t) = |E(t)|\exp\{i\Phi(t)\}\exp\{i\Phi_0\}\exp\{-i\omega_0 t\}$$

where $|E(t)|$ is the time-dependent envelope, which is related to the de Broglie wavelength distribution in space. ω_0 is the carrier frequency (usually the angular frequency of the vorticity photon in the corpuscle electron's internal frame at rest state). $\Phi(t)$ is the time-dependent phase, and Φ_0 is a constant, called the "carrier-envelope offset phase." The square of the envelope $I = |E(t)|^2$ is the time-dependent instantaneous power of the corpuscle electron, which can be measured in an electron microscope if a detector (e.g., a CCD camera or photographic film) with proper bandwidth is available. The derivative of the time-dependent phase accounts for the occurrence of different frequencies at different times, i.e., $\varphi(t) = \partial\Phi(t)/\partial t$ is the instantaneous frequency of the pulse of an electron that describes the oscillations of the electric field around that time.

If using a representation of the corpuscle electron in the chronocyclic phase space, we should take a Fourier transformation of the correlation function corresponding to the time difference of the two electric fields, such as

$$W(t, \omega) = \int dt' \, \{E(t + (1/2)t')E^*(t - (1/2)t')\}\exp(i\omega t')$$

or

$$W(t, \omega) = \int (d\omega'/2\pi)\{\check{E}(\omega + (1/2)\omega')\check{E}^*(\omega - (1/2)\omega')\}\exp(-i\omega' t)$$

where Ě and Ě* are the Fourier transforms of the electric field E.

The function *W* is called the "chronocyclic" Wigner function. Useful features of the Wigner function are that it is real-valued, and its marginals are the temporal and spectral intensities

$$I(t) = |E(t)|^2 = \int (d\omega/2\pi)W(t, \omega)$$

$$\check{I}(\omega) = |\check{E}(\omega)|^2 = \int dt W(t, \omega)$$

The Wigner function is sufficient to characterize both individual corpuscles and partially coherent corpuscle ensembles. It is not, in general, positive definite and cannot be considered a probability distribution of the electron field. Indeed, negative Wigner functions are quite common even for simple corpuscle shapes and also characterize many of the complicated corpuscle electron shapes. In an electron microscope, the only measurable physical parameter is the electron's energy at the focus and image plane. The electron energy loss spectrum actually analyzes the velocities corresponding to the frequency of the rotating vorticity electromagnetic fields for inelastically scattered electrons. It also measures the frequencies or energy. Understanding these relations is very important for electron microscopy.

2.3 Electrons flying through a magnetic lens

Again, a corpuscle electron can be expressed as displaying the ψ function:

$$\Psi_{db} = \Psi_0(r, p, t)\exp[i/h(pr - \varepsilon_v t)]$$

where Ψ_{db} is the de Broglie wave function and expresses the spatiotemporal variation characteristics of a flying electron. The changeable energy of the electron is its kinetic energy, ε_v, while it is moving. The transient time is $\delta t = d\ell/v_t$, while it travels on a trajectory in space. The x, y, z, P_x, P_y, P_z are positions and corresponding momentum in the coordinate system of the laboratory frame. The velocity of the electron has to be obtained from Schrödinger's equation, which is derived from the laws of Maxwell and Newton.

Because $|\Psi(x, y, z, t)|^2 = \rho_{em}(x, y, z)$ or ρ_m, and the current of mass or electric charge along the x dimension is

$$J(x,t) = i\frac{\hbar}{2m}\left(\psi(x,t)\frac{\partial\psi^*(x,t)}{\partial x} - \psi^*(x,t)\frac{\partial\psi(x,t)}{\partial x}\right)$$

And $J(x) = v_t\ \rho_{em}(x)$, or $v_t\ \rho_m(x)$. Therefore, the velocity of an electron on the trajectory is

$$v(x,t) = \frac{J(x,t)}{|\psi(x,t)|^2}$$

The electron's velocity can be obtained from the electron's current and electric density (or amplitude). For free motion of the electron, the spatial interval between two electrons should be larger than the dimension of a corpuscle electron, which is the effective wave length of the de Broglie wave, $\lambda_{db} = \lambda c/v_t = h/\delta p$, which emanating from Heisenberg uncertainty. It is clear that the velocity is a vector, which has different components in a chosen coordinate system. If the longitudinal direction is the optic axis in an electron microscope and the transverse direction is in the specimen's plane, then the velocity of the moving electron along the longitudinal direction (usually chosen as the z direction for the optic axis) is tremendously larger than the transverse one, which is perpendicular to the optic axis or z direction. Therefore, the electrons in an electron microscope are almost, one by one, sequentially flying with given longitudinal intervals (see Table 2.1), and the electromagnetic lens of the electron microscope can impart the ratio between longitudinal and transverse velocity components during the electron's passage through the magnetic field of the lens (≈ nanosecond or ns) due to the Lorentz force. This induces the trajectory of an electron in an electron microscope to be helical curves with different ratios of transverse and longitudinal velocities. The effective wave length of the de Broglie wave gets either shorter while the transverse velocity increases or longer while the transverse velocity decreases. The ratio of transverse and longitudinal velocities is also related to the divergence of the helical motion of the electron, which is influenced by the emitter and the gun characteristics of the microscope. The degree of divergence dictates the helical diameter of the flying electron. While transverse velocity is falling, the transverse wave length is increasing, which implies the transverse occupied dimension of the corpuscle electron is larger. For example, if the transverse dimension of the corpuscle electron is 1 mm, the transverse velocity is about $\approx 10^{-4}$ m/s or 10^{-6} nm/ns. This might suggest that the electron is almost sequentially flying, as shown in Figure 2.3.

The stationary electric field accelerates the velocity of the electron along the symmetrically distributed electric flux between the anode and cathode of an electron gun. The velocity of the electron at the anode plane is almost parallel to the optic axis of the microscope, as shown in Figure 2.3.

Focus of the static electric lens (i.e., electron gun) is the crossover point of electron trajectories, and different electrons cross over the optic axis with divergent angles, which may be $\alpha \approx 10^{-3}$ rad, at different times. However, the image of the electron

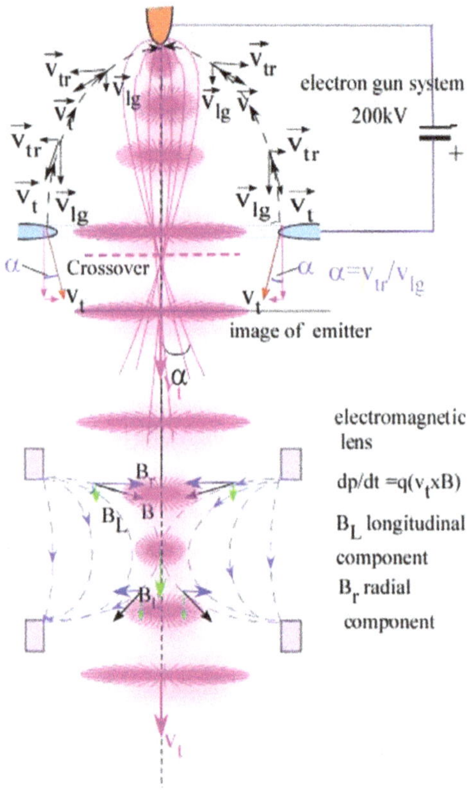

Figure 2.3: Dimensional modification of an electron when it is moving through an electron gun and an electromagnetic lens.

emitter is located in the image plane of the electron gun, which is below the crossover point. For electron microscopy, the crossover plane is more useful than the image of the emitter. The divergence angle around the crossover point dominates the distribution dimension of the electrons flying out of the anode on the crossover plane.

While an electron is passing through an electromagnetic lens, the magnetic field modifies the velocity of the corpuscle electron, causing it to vary. The transverse velocity of the corpuscle electron increases or decreases at the upper or lower part of the center principal plane of the lens. It also instantaneously modifies the longitudinal flying velocity due to Maupertuis's law. Because the total kinetic energy of the corpuscle electron is conserved, when the transverse velocity of the electron increases (or decreases), the longitudinal velocity instantaneously decreases (or increases). Therefore, the de Broglie wave length along the direction of flight may be adjusted before and after the magnetic lens, but the total energy should remain the same. The passing period through the electromagnetic lens is approximately 2–3 × 10^{-11} s (or 0.2–0.3 ps) for a general electron microscope, and the action of the magnetic field may influence

Table 2.1: Basic data of an electron in an electron microscope.

Accelerating voltage (kV)	Relativistic wavelength (nm)	Mass (x m_0)	Velocity (x 3 × 10^8m/s)	Interval of two electrons in succession along the velocity direction	
				I PA (Picoamp)	100 PA
100	0.00370	1.196	0.54	27 m	27 mm
200	0.00251	1.391	0.69	34 m	34 mm
300	0.00197	1.587	0.78	39 m	39 mm
400	0.00164	1.783	0.83	46 m	46 mm
1,000	0.00087	2.957	0.94	47 m	47 mm

the path of the electron's trajectory, but the total kinetic energy does not increase. In other words, the ratio of transverse and longitudinal momentum can be manipulated by the magnetic lens, resulting in a varying convergent angle of an electron and the dimension of the de Broglie wave length. When the kinetic energy of a corpuscle electron is concentrated in the transverse velocity direction, the de Broglie wave length is shorter, but when the kinetic energy is concentrated in the longitudinal velocity direction, the de Broglie wave length in the transverse direction is longer. Using a magnetic lens to steer the ratio of the transverse and longitudinal momentum is an important technique in electron microscopy, especially for so-called high-resolution electron microscopy. Figure 2.4 demonstrates that the center principal planes of the magnetic lens have a minimal transverse wave length with a longer longitudinal de Broglie wave length. The location of a sample in a transmission electron microscope is near the front focal plane of the objective lens. The velocity divergence of the corpuscle electron exiting from the sample can be sorted out by the lens to guarantee the velocity orientation of the exited electrons from the sample.

The upper part of the lens acts as a convergent for the divergent velocity of particles, and the downward part of the lens manipulates the outgoing trajectories of the electrons.

This is significant for image formation in an electron microscope. The downward part of the magnetic lens dominates the back focus position. In other words, the outgoing path of the trajectory would form a cross-point with the optical axis that should be the back-focus point of the magnetic lens. This back focus builds up the back-focus plane of the objective lens in a transmission electron microscope. Because of the sample being located near the front focus plane of the objective lens, the transversal magnification would be $M_T = (1/\delta)f$ where f is the front focus length and δ is the deviation distance between the front focus and the plane of the sample. If δ is zero, the transversal magnification will be infinitely large.

The defocus of the magnetic lens means varying the cross-point location of the outgoing trajectory with the optic axis, which implies the variation of the phase of the

$$\frac{dp}{dt} = e(v_t \times B(t,r))$$
$$= e(v_r \times B_z + v_z \times B_r)$$

$$\tau = l/v_t \qquad \int_{t=0}^{t=\tau} \frac{dp}{dt} dt = p_0 - p_\tau$$

(a)

(b)

(c)

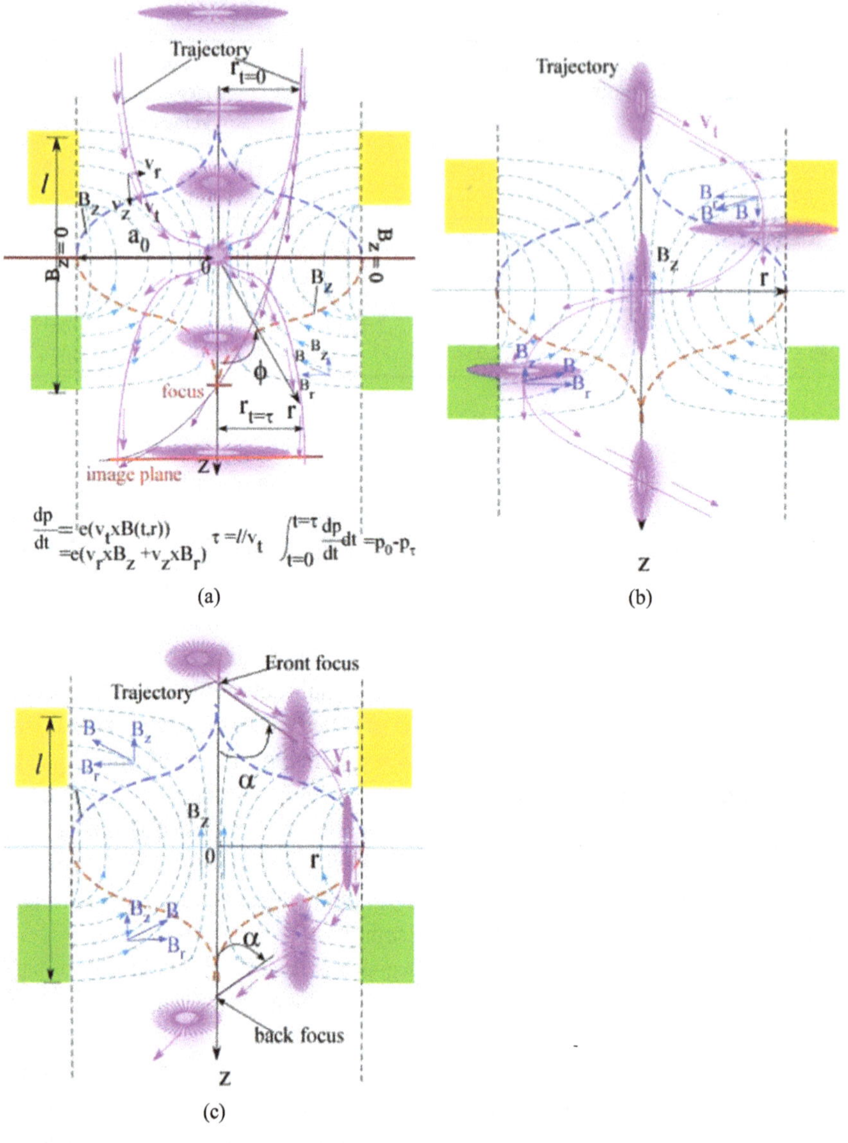

Figure 2.4: The velocities of the different trajectories and the variation of the de Broglie wave length in transverse (r) and longitudinal (z) orientation. (a) Parallel incoming, (b) inclined incoming with crossing at the center of the lens, and (c) inclined through the front focus and out through the back focus.

electromagnetic waves in the corpuscle electron. Detecting an electron involves measuring the electric energy or charge by fluorescent atoms or a CCD device, and the mass velocity spectrum in the laboratory frame. The electric field energy of an electron can be expressed as

$$E_r = E_0\exp\{-i[\omega_0 t - (\omega r/c)\theta]\} = E_0\exp\{-i[\omega_0 t - k_{(r\theta)}\theta]\} = \Psi_{db}$$

The energy of an electron is $|E_r|^2 = E_0{}^2 = |\Psi_{db}|^2$. The electric field distribution is dominated by the phase factor, $k_{(r\theta)}\theta$, at the coordinate points, in which the de Broglie wave number $k_{(r\theta)}$ multiplies the rotating angle. However, the energy is related to the frequency of the rotating electromagnetic vorticity photon, which varies with the velocity of the corpuscle electron.

Electron microscopy cannot analyze the velocity variation parameters yet. The phase factors of the corpuscle electron may be measured by Planck's constant, h. Therefore, if the phase factor is tremendously larger than Planck's constant (usually called far-field), the effect on the phase of the corpuscle electron is difficult to observe in the laboratory frame. For transmission electron microscopes, the magnification can be $\times 10^6$–10^8 and δx may be, at scale, about $\times 10^{-6}$–10^{-8} mm or ≈ nm. de Broglie transverse wave length may approach the same dimension by adjusting the magnetic lens, as shown earlier. This provides the possibility to observe the assembly of repartitioned electron density distribution from exited samples located near the front focus plane (near field), which is the Talbot effect. The velocity spatial divergence of the corpuscle electron induced by the periodic potential of the crystal, called a "phase grating," can be observed at the back focus plane as diffraction patterns and using an objective aperture to collect the high-resolution image. The electron microscope cannot collect the phase variation of the corpuscle electrons except by using the strong standing waves of a strong laser optic field with high frequency, which is the "Kapitza-Dirac effect."

2.4 Characteristics of the confined electromagnetic wave energy flux of a corpuscle electron

If we accept that a corpuscle electron is the confined electromagnetic wave energy flux of a vorticity photon, then an electron is a corpuscle of confined electromagnetic energy that has its own electromagnetic mass and follows Newton's second law. The electromagnetic mass emanates from the frequency of the photon with an S(2) × S(2) dimensional circulation at the speed of light in its internal frame. The circulating frequency is about $\nu_0 = m_0c^2/\hbar$, above 10^{22} Hz, or a few zetahertz (ZHz). The frequency is super-high, implying that it is impossible to resolve the single periodicity in the laboratory frame. The electric charge is usually the measurable characteristic parameter of this confined electromagnetic energy in the laboratory frame. Maxwell's equations would control the dynamic propagation of the electromagnetic fields during free flight in the laboratory frame.

The stretching dimension of the circulating photon's tracks has three dimensions. The momentum or electromagnetic wave vector (Poynting vector) must change its orientation in a period of time (≈10^{-21} s). The total electromagnetic energy and kinetic energy of the circulating photon in the volume of a corpuscle electron should be conserved as

$$\frac{d}{dt}(E + T) = 0$$

and the kinetic momentum of the circulating photon and its electromagnetic momentum (the Poynting vector) are also conserved as

$$-\frac{d}{dt}\left(\vec{p}_i + \int d^3r \frac{1}{c}(E \times B)_i\right) = 0$$

in which

$$\int d^3r \frac{1}{c}(E \times B)$$

is the electromagnetic momentum of the volume of a corpuscle electron. If the density of the electromagnetic momentum is expressed as

$$\vec{G} = \frac{1}{c}(E \times B)$$

Its rotation creates angular momentum by rotating around the center of the electromagnetic field energy. The center of energy coordination of the circulating electromagnetic field energy may present as

$$R(t) \equiv \frac{\int \vec{r}\, U(r,t) d^3r}{\int U(r,t) d^3r}$$

where U(r,t) is the loci and elapsed energy flux of vorticity in the electromagnetic field, and **r** is the radiated distance between the center and the elapsed energy flux. At rest condition, this center would be the origin point of the internal coordinate frame of the corpuscle electron, which means $R = 0$.

In this condition, the electromagnetic energy is constant. The distribution of electromagnetic energy flux holds an effective mass as $m_0 = \varepsilon_{em}/c^2 = h\omega_0$. The dimension metric scale is $\lambda_c = \hbar/mc$, which is the Compton wave length, and the proper time scale $\tau = \lambda_c/c = \hbar/mc^2$, which means time periodicity for circulating a circle circumference with a radius $\lambda_c = \hbar/mc$. It is clear that the corpuscle electron is a confined electromagnetic energy flux assembled in the spatiotemporal contours of the corpuscle electron.

The center of energy coordination of the circulating electromagnetic field is the center of the spatial volume of a corpuscle electron. The velocity of the corpuscle electron is the moving speed of the center of energy coordination of the confined electromagnetic field flux. The distribution or spreading dimension of the confined circulating electromagnetic field is the shape of the corpuscle electron. Because of the relativity feature between the rest (or at its internal frame) and motion frames as the laboratory frame, the images seen on the screen of an electron microscope should be the image of the collected corpuscle electrons at low magnification ($<\times 10^5$) in the laboratory frame, but for higher magnification ($>\times 10^5$), the observed image may be the assembly of the electromagnetic energy carriers, which is the corpuscle electron. Each corpuscle is steered by the phase velocity of the circulating photons in the internal frame of the corpuscle electron. In cases of low magnification, the mass and charge with spin are used for electron optics, which are based on the Maxwell and Newton equations. Thickness-density contrast is based on the particles' scattering and absorption. For higher magnification, the distribution and dynamic process of the confined single harmonic electromagnetic wave of the photon in the internal frame have to be considered. The kinetic characteristics around the center of the energy coordination of the confined single harmonic electromagnetic wave of the photon and the phase of the confined single harmonic electromagnetic wave (i.e., the group and phase velocity) have to be manipulated by Maxwell and Newton laws. Maxwell's equations control the electromagnetic field momentum propagation (i.e., phase velocity), leading to the velocity variation at the center of the energy coordination of the electromagnetic energy flux (i.e., group velocity) by de Broglie relation, $\mathbf{v}_g \bullet \mathbf{v}_{ph} = c^2$.

Based on these intrinsic characteristics of a corpuscle electron, the distribution of the circulating electromagnetic field in space may be expressed as an enveloped function. In the laboratory frame, the frequency of the confined single harmonic electromagnetic wave of the photon will be $\nu = \nu_0(1-(v_t/c)^2)^{-1/2}$, which may be expressed as $\nu = [\nu_0 + 1/2(v_t/c)\nu_0 + 3/8(v_t/c)^2\nu_0] = \nu_0 + \delta\nu_0$, where $\delta\nu_0 = [1/2(v_t/c) + 3/8(v_t/c)^2]\nu_0$.$\delta\nu_0$ is the frequency of the de Broglie wave. Due to $\omega = 2\pi\nu$ and $\omega = [1 + 1/2(v_t/c) + 3/8(v_t/c)^2]\omega_0$, the frequency or angular frequency is a function of the velocity of the corpuscle electron. However, the trajectory of a corpuscle electron is not always linear and usually has curved lines in the laboratory frame. When the corpuscle electron passes through a magnetic lens or atoms in a sample, the magnitudes and direction of the velocity are time- or location-dependent on corresponding trajectories based on the Maupertuis law. The frequency of the confined electromagnetic field flux in the internal frame of a moving corpuscle electron would be a time-dependent function in which a carrier frequency, ν_0 or ω_0, determined by the energy, $\varepsilon_0 = m_0c^2 = \hbar\nu_0$, is extremely high (ZHz), and the additional frequency, $\delta\,\nu_0$ or $\delta\omega_0$, emanating from the kinetic motion of the corpuscle electron may be lower, then it may be matched with the atom's vibration and spatial distribution. Because the kinetic motion can vary with time and orientation of moving velocity, the electric field of a corpuscle electron, E, can be expressed as

$$\mathbf{E}(x,y,z,t) = E(x,y,z,t)\exp[i(\omega t - \mathbf{k}_{(r\theta)}\theta)] = E(x,y,z,t)\exp[i(\omega_0 t + \delta\omega_0 t - \mathbf{k}_{(r\theta)}\theta)]$$

The slowly varying envelope function is $E(x,y,z,t)\exp[i\omega_0 t]$, and the kinetic motion inducing the variation may show as $\exp[i(\delta\omega_0 t - \mathbf{k}_{(r\theta)}\theta)]$. If we could see the envelope function in frequency space, it would be $(x,y,z,\omega) = E_0\delta(\omega-\omega_0)$, where $\delta(\omega\text{-}\omega_0)$ is the delta function and electromagnetic energy of the corpuscle electron is $\varepsilon = |E_0|^2$. The distribution of the electromagnetic field is steered by the phase of circulating electromagnetic field energy flux as $\exp[i(\delta\omega_0 t - \mathbf{k}_{(r\theta)}\theta)]$, in which the kinetic energy of the corpuscle electron induces an additional frequency, $\delta\omega_0$, which depends on the direction and magnitude of the corpuscle electron's velocity such that $\delta\omega_0 = [1/2(\mathbf{v}_t/c) + 3/8(\mathbf{v}_t/c)^2]\omega_0$. The phase velocity, $\delta v_{ph} = \delta(\omega_0/k_{(r\theta)})$, is also related to the direction and magnitude of the velocity of the corpuscle electron. Therefore, in frequency space, the energy variation is sorted out by the frequency spectrum. Because we believe the corpuscle electron is the confined rotation of a circulating photon, the kinetic energy of a traveling photon around the closed curl curves stores the electromagnetic energy in space and time. The trajectory corresponding to spatial and temporal variations is a ray of the corpuscle electron that results from the elapsed time and velocity of the corpuscle electron in the laboratory frame. However, the elapsed time and velocity at any location of the corpuscle electron's ray are not the same for the internal dynamic motion of the corpuscle electron, based on the relativity principle of the inertial frame. The spatial and temporal velocity relationship between laboratory and internal frames can be expressed as the de Broglie relation $v_g \bullet v_{ph} = c^2$ and Einstein relation, $\varepsilon = \hbar v = h\omega$.

Based on the relationship between the wave length and periodicity, the phase velocity of the de Broglie wave is related to the velocity of the corpuscle electron, v_t. If the time periodicity of the de Broglie wave is T_{db}, then

$$\lambda_{db} = v_t T_{db} = v_t/\omega_{db} = v_t(h/\varepsilon_{db}) = v_t(h/mc^2) = (v_t/c)(h/mc) = (v_t/c)\lambda_c$$

where λ_c is the Compton wave length corresponding to the circumference contours of the circulating vorticity field of a photon. These formulas indicate that the de Broglie wave of a corpuscle electron is different from the Compton wave length and that the de Broglie wave of an electron is a modified kinetic process of the circulating vorticity field of the photon. Therefore, understanding the intrinsic characteristics of a corpuscle electron requires understanding the dynamic motion duality of both particle and wave. The charged particle feature of a corpuscle electron emanates from the circulating vorticity field of the photon with a frequency of a few ZHz at dimension $\lambda = h/mc$ (where $m = m_0(1-(v_t/c)^2)^{-1/2}$) in the laboratory frame). Because $\lambda = h/mc$ is around 10^{-12} m, and human eyes can only resolve the difference in distance at around 10^{-4} m, the charged corpuscle electron is an energetic, charged particle in the laboratory frame or in an electron microscope. The electromagnetic wave characteristics of the corpuscle electron exhibit particular interactions between the external and internal electromagnetic fields only at its Compton wave length scale and induce frequency modification and phase velocity variations, leading to interference in the spatial and temporal domains of the corpuscle electron, which is a near-field situation in its internal frame. This indicates that intensity induced by phase may be observed only in the near field, and Fraunhofer diffraction is a far-field situation, which cannot observe intensity induced by phase in the laboratory frame. de Broglie and Einstein relationships, $\mathbf{v}_t \cdot \mathbf{v}_{ph} = c^2$, and $\mathcal{E} = h\omega = mc^2$, decipher the nature of the electron wave as particle velocity and phase wave velocity.

References

Elbaz, C. "Wave-particle duality in Einstein-de Broglie programs." Journal of Modern Physics. 5, 2192–2199, (2014).

Hamdan, N. "The dynamical de Broglie theory." Annales de la Fondation Louis de Broglie. 32(1), 11, (2007).

Hill, J.M. "Combining Newton's second law and de Broglie's particle-wave duality." Results in Physics. 8, 121–127, (2018).

Logiurato, F. "Relativistic derivations of de Broglie and Planck-Einstein equations." Journal of Modern Physics. 5, 1–7, (2014).

Lush, D.C. "Similarity of the Doppler shifted time-symmetric electromagnetic field of a Dirac particle to the de Broglie matter wave."aiXiv: 1609.04446v2 [physics-classic=ph] (9 Oct 2016).

Marquet, P. "On the physical nature of the de Broglie wave." Progress in Physics. 12, 318, (2016).

Rose, H. "Optics of high-performance electron microscopes." Science and Technology of Advanced Materials. 9 (2008), 01401071.

Vankov, A. "On de Broglie wave nature." Annales de la Fondation Louis de Broglie. 30(1), 15, (2005).

Chapter 3
Motion duality of electron in space and time

3.1 The electron flying in electron microscope

It may be valuable to think that flying electrons with a given velocity starts as nearly static corpuscles, is accelerated in an electron gun, and then escapes from the electron gun to free-fly in the column of an electron microscope and through several magnetic and/or electric lenses and the sample, finally arriving at a detector (fluorescent screen or charge-coupled device [CCD]). As we discussed earlier, an electron is an energetic particle in the laboratory frame and an ensemble of tracks of a circulating vorticity photon with the speed of light in its internal frame. In short, it is a quantum particle and wave. In the laboratory frame, as in an electron microscope, the flying electron traces form the paths of the electron near the optic axis. Due to the fact that electrons have charge (as -q) and if their flying velocity is v_t, the flying electrons create the electric current I_e (as mentioned $I_e \approx 10^2 \sim 10^3$ picoamperes in a usual electron microscope), but two electrons cannot get close to each other due to their anti-bunch property of charged particles in the laboratory frame. However, the electric charge of an electron is the result of the circulating electromagnetic vorticity field of a dynamically moving photon in the electron's internal frame. So any spatio-temporal points (x_i, y_i, z_i (=$v_t t_i$)) on the trajectory of the paths of a flying electron in the laboratory frame must simultaneously have the corresponding dynamic parameters (x', y', z'(=v_t' t')) of the circulating vorticity photon in the electron's internal frame. In other words, the kinetic parameters of the flying electrons in the laboratory' four-dimensional space-time (Minkowski space-time) have to connect to the dynamic parameters of the circulating vorticity photon in the electron's internal four-dimensional space-time. In the laboratory frame, an electron is a linearly flying charged particle and simultaneously has its internal dynamic circulating motion of the photon. This motion duality of an electron is the duality of particle and wave of the electron. Understanding this motion duality is essential for transmission electron microscopy (TEM).

Electron traveling in an electron microscope has to be understood as a space-time process. The concept of *space-time* is an important physical concept that has fused the three dimensions of space and the one dimension of time into a single four-dimensional continuum called four-dimensional Minkowski space-time. The causality of any motion events has to be involved, and the relationship between the events occurring at any spatial location and their corresponding time has been exposed. For example, the wavefront of the de Broglie wave is the ensemble of the same dynamic states of the harmonic oscillator, which is the circulating photon, at its internal space with simultaneity. The wavefront of the de Broglie wave of an electron is an important physical parameter for electron microscopy. These parameters in the laboratory frame are the parameters transformed by Lorentz transformation from the electron's internal frame.

https://doi.org/10.1515/9783111449333-003

If the parameters in the laboratory frame are unprimed letters and the primed letters show the physical parameters in the corpuscle electron frame, which is flying with velocity v_t along the z-direction (which is the optic axis of the electron microscope) in the laboratory frame, then we would have the following relations between these two frames:

$$z = \gamma(z' + v_t t'); \; ct = \gamma[ct' + (v_t/c)z']:$$

$$z' = \gamma[z - (v_t/c)ct]; \; ct' = \gamma[ct - (v_t/c)z]$$

$$k_z = \gamma[k'_z + (v_t/c)(\omega'/c)]; \omega/c = \gamma[\omega'/c + (v_t/c)k'_z]$$

$$k'_z = \gamma[k_z - (v_t/c)(\omega/c)]; \; \omega'/c = \gamma[\omega/c - (v_t/c)k_z]$$

$$E'_x = \gamma(E_x - v_t B_y); \; E'_y = \gamma(E_y + v_t B_x); \; E'_z = E_z$$

$$B'_x = \gamma(B_x + (v_t/c^2)E_y); \; B'_y = \gamma(B_y - (v_t/c^2)E_x); \; B'_z = B_z$$

If $\mathbf{E}_{||}$; $\mathbf{B}_{||}$ and $\mathbf{E}_{\perp}$; $\mathbf{B}_{\perp}$ are the components of the fields, which are parallel and perpendicular to the velocity $\mathbf{v}$, then these parameters would be determined by the Lorentz transformation as

$$\mathbf{E}'_{||} = \mathbf{E}_{||} \; \mathbf{B}'_{||} = \mathbf{B}_{||}$$

$$\mathbf{E}'_{\perp} = (\mathbf{E}_{\perp} + \mathbf{v}x\mathbf{B}_{\perp})/(1 - (\mathbf{v}/\mathbf{c})^2)^{1/2} \; \mathbf{B}'_{\perp} = (\mathbf{E}_{\perp} - \mathbf{v}x\mathbf{B}_{\perp})/(1 - (\mathbf{v}/\mathbf{c})^2)^{1/2}$$

These equations show that under Lorentz transformations, electric and magnetic fields in the laboratory frame will convert into the fields in the electron's internal frame.

Based on relativistic principles, the phase of the de Broglie wave would be the same in these two frames:

$$\varphi = \varphi' \;\; \text{or} \;\; k_z z - \omega t = k_z' z' - \omega' t'$$

The flying corpuscle electron at any elapsed time and locus on the trajectory in the electron microscope frame has a charge with a corresponding confined dynamic electromagnetic vorticity field in its own internal space-time. The physical parameters such as v_t, ω_t, k_t, r_t, $\mathcal{E}_t$, $\mathbf{E}$, $\mathbf{B}$, ϵ, and μ would be found and measured in the electron microscope frame, but they are related to the corresponding values in its own internal frame.

The corpuscle electron, as an energy carrier with the electromagnetic field frequency, $\nu_0 = m_0 c^2/\hbar$, is accelerated to a certain velocity v_t to escape the electron gun. Its total energy approaches $\varepsilon = \hbar\nu = \hbar\nu_0/(1 - (v_t/c)^2)^{1/2}$ indicating that the flying velocity v_t can modify the frequency of the local spatial field where it passes through. The frequency of the confined electromagnetic field in the electron's internal frame would also vary with traveling time and spatial locus. The relationship between frequency (or time) and spatial locus would be the codes of the interaction process on the trajectory of the flying electron. Therefore, exploiting spatial and temporal features of the interactions during electron flight would be the target of electron microscopy.

As well known, the velocity of a corpuscle electron is a group velocity, or the velocity of the energy center of a corpuscle electron, and may be expressed as

$$v_t = v_g = d\omega/dk = (d\omega/d\varepsilon)(d\varepsilon/dk) = 1/h\,(d\varepsilon/dk)$$

Because $\varepsilon = \hbar\omega = \hbar\nu$ and $d\varepsilon = \hbar d\nu = \hbar\nu_0 d[1 + 1/2(v_t/c) + 3/8(v_t/c)^2] = \varepsilon_0 df(v_t/c)$, and where $f(v_t/c) = [1 + 1/2(v_t/c) + 3/8(v_t/c)^2]$, and using the de Broglie relation, $\mathbf{v}_t\mathbf{v}_{ph} = c^2$, that $v_t = (1/h)d\varepsilon/dk = \varepsilon_0 df(c/\mathbf{v}_{ph})dk$ implies the velocity of the corpuscle electron relates to the function of phase velocity varying with the phase wave vector changing in its own frame. The phase of circulation of the electromagnetic vorticity field of the photon in the internal frame of the corpuscle electron may be written as

$$d\varphi = 2\pi[\nu t - (\nu/v_t)ds] = 2\pi[(\nu/v_t)v_t t) - (\nu/v_t)ds] = 2\pi[\nu/v_t(v_t t - ds)]$$

$$v_{ph} = d\varphi/dt = 2\pi[\nu/v_t(v_t - ds/dt)] = 2\pi[\nu/v_t(v_t - r d\theta/dt)]$$

The corpuscle electron's velocity v_t is closely related to the confined circulating electromagnetic energy's rotating velocity ($d\theta/dt$), and the ratio (ν/v_t) of the wave frequency (ν) and velocity (v_t) of the center, around which the confined electromagnetic vorticity field of the photon circulates, and the difference between the group velocity (v_t) and the transient circulating velocity ($r d\theta/dt$) of the photon. That means (1) the ($v_t - r d\theta/dt$) manipulates the phase variation with time and (2) if the velocity v_t of the electron in the laboratory frame corresponds to the elapsed path ds on the trajectory of the circulating photon, then the phase is constant. Therefore, the motion of the electron as a particle in the laboratory frame is intimately related to the internal periodic kinetic motion of the corpuscle electron, which is the particle-wave duality of the electron.

3.2 Electron beam and a corpuscle electron

The corpuscle electron holds an electric field, emanated from the circulation of the electromagnetic vorticity field of the photon, but in the laboratory frame, the electrons are anti-bunched, which means two electrons cannot occupy the same location in space due to the repelling effect of their electric fields. The flying electron in the vacuum of an electron microscope follows Maxwell's law, which means parallel flying electrons should repel each other. Therefore, electrons should fly consecutively. The consecutive flying electrons form the electron beam. The current of the beam is dominated by the interval between two consecutive flying electrons with a definite spatial and temporal interval, which depends on the emission characteristics of the tip of the electron gun. If the electron beam is visualized as a wave, then the wavefront of the electron beam should be an assembly of the same kinetic state (or momentum) with simultaneity.

However, the emission process of the tip of the electron gun is statistical rather than periodic. Actually, the electron is not only a particle but also a particle and a wave. In other words, an electron has its own space-time frame, and the energy and momentum of an electron originate from the circulation of the electromagnetic vorticity field of the photon, as discussed in Chapter 1. The electron beam describes the assembly of the paths of trajectories of flying corpuscle electrons, in which the electromagnetic vorticity field has dynamic motion with circular angular velocity 2ω and hyperbolic angular velocity η/c in the internal space-time frame. The dynamic motion may be imagined, as discussed in Chapter 1, as two perpendicular two-dimensional circulations building up a third motion in its internal frame (or corpuscle contour), which is seen as linear or curvilinear motion in the laboratory frame. From this view, we simply suppose that any velocity of a flying electron can be decomposed into two components: one parallel to the optic axis of the electron microscope as $v_{\parallel}$ and the other perpendicular to the optic axis of the electron microscope as $v_{\perp}$. Using these velocity components, we may connect the velocity in the microscope frame with the corresponding kinetic motion of the electromagnetic vorticity field of the photon in its own frame without considering the detailed process. Particularly, we may focus on the carrying energy, momentum, and phase, which relate to the circulation of the electromagnetic vorticity field of the photon.

Usually, describing an electron beam as an electron wave may not be valuable because, at the microscope frame, the electron beam is an energetic flow of charged particles. Though helical, sequentially flying electrons may have mechanical rotational angular momentum, it is not related to the circulation of the electromagnetic vorticity field of the photon with energy $\mathcal{E} = h\omega$. Our understanding of the electron wave is the circulating electromagnetic vorticity field of the photon in an electron's internal space-time. Its metric scale is the Planck constant, and its dynamic activity is confined within the Compton wavelength, which may be considered the contour of a corpuscle. That is its nature of duality of particle and wave.

If an electron is a confined electromagnetic vorticity field of the dynamic photon in corpuscle contours at its internal frame and electron microscope frame, as discussed in Chapter 1, a photon may be an electromagnetic field vortex with energy $\mathcal{E} = h\omega$. The energy of a dynamic photon is the accumulation of instantaneous electromagnetic vorticity energy:

$$\varepsilon = \int \varepsilon(\theta', r', t')d\theta' dr' dt' = h\sum_{i}\int \omega_i(\theta', r', t)d\theta' dr' dt'$$

The electromagnetic vorticity energy stems from vorticity in the electric and magnetic fields:

$$\partial E(r', t')/\partial t' = (1/\mu_0\varepsilon_0)\nabla \times \mathbf{H}, \quad \partial B(r, t)/\partial t = -\nabla \times \mathbf{E}$$

That means the vorticity velocity field of the magnetic vector potential at spatial contours is related to the velocity of the electric field, corresponding to the temporal metric scale. Simultaneously, the vorticity gradient field of the electric potential at spatial contours induces a magnetic velocity field. The element electromagnetic energy density is as

$$d\varepsilon_i = {}_0c^2\nabla\cdot(\mathbf{H}\times\mathbf{E}) + \{\varepsilon_0c^2\mathbf{H}\cdot(\nabla\times\mathbf{E}) - \partial(1/\mu_0\varepsilon_0)(\nabla\times\mathrm{H})\cdot\mathbf{E}))\}$$

That implies the energy of the electric and magnetic vorticity field such as

$$\varepsilon_{vort} = \{\varepsilon_0c^2\mathbf{H}\cdot(\nabla\times\mathbf{E}) - \partial(c^2(\nabla\times\mathrm{H})\cdot\mathbf{E}))\}$$

It is part of the total energy and relates to the dynamic energy of the vorticity electric and magnetic fields, such as $\varepsilon_{dyn} = \epsilon_0c^2\nabla\cdot(\mathbf{H}\times\mathbf{E})$ what is called the Poynting dynamic energy. $\mathrm{P} = \varepsilon_{dyn}/\mathrm{c} = \epsilon_0\mathrm{c}\nabla\cdot(\mathbf{H}\times\mathbf{E})$ is the momentum of the electromagnetic vorticity field called as the Poynting vector. These relations tell us that the electromagnetic energy and momentum density may be balanced in a confined spatial contour (or element volume), which means there isn't any energy exchange with the exterior, but the electromagnetic momentum flux circulates with circling time, τ, in this spatial contour, resulting in a radiating electric field as an electron charge at the laboratory frame, as shown in Figure 1.8. It is clear that space-time metrics are essential parameters. At the electron microscope frame, the electron beam is a charge flow or electric field flow with a vorticity magnetic field in a vacuum. Because the radiated electric field strength of the flying electron varies along the radiate direction, the flying velocity results in the electromagnetic energy density and momentum (Poynting) flux parallel and perpendicular to the velocity direction having different distribution functions $\widehat{\mathrm{E}}_{\mathrm{L}}(\mathrm{E_r}, \mathrm{H}_\omega) = \widehat{\mathrm{E}}_{\mathrm{L}}(\mathrm{r}, \omega)$. This electromagnetic field is related to electromagnetic fields in the corpuscle electron, but it is not the same because the flying electron is moving in the laboratory frame, while the circulating electromagnetic vorticity field is an electromagnetic dynamic momentum synthesized motion with two two-dimensional circulations. The nature of the physical entity of an electron is to carry the confined electromagnetic energy in space and dynamic electromagnetic energy flux as electromagnetic momentum density flux.

3.3 Energy and momentum and space-time

As is well known, TEM usually manipulates the beam orientation for different modes such as tilting, converging, and scanning. As mentioned before, TEM fundamentally steers the direction of velocity of the corpuscle electron to manipulate the internal kinetic and dynamic motion of the circulating electromagnetic vorticity field of the photon that induces the different actions in the observed sample.

The temporal duration of a corpuscle electron traveling through the sample is $\tau_z = \ell/v_z$, where ℓ is the thickness of the sample and τ_z is the time of passing through the sample. During the temporal range τ_z the sample and corpuscle electron overlap each other. This temporal period is determined by the longitudinal (on the optical axis) component ($v_{\parallel}$) of the velocity of the corpuscle electron, and the velocity ($v_{\perp}$) perpendicular to the optical axis, or transversal velocity, will have small energy based on the magnitude of the velocity, which is steered by the inclined angle between the velocity and optic axis. Therefore, during the overlapped temporal duration τ_z of the electron corpuscle with the sample, the dynamic electromagnetic field of the circulating photon in the corpuscle electron's internal frame has to be in the electromagnetic environment of the sample, which is different from a vacuum. The overlapped temporal duration dominates the electromagnetic interaction process that modifies the dynamic motions of the circulating photon in the corpuscle electron, resulting in the kinetic behaviors of the escaped corpuscle electron from the sample.

The flying corpuscle electron's velocity direction is the electromagnetic energy flux direction, but the electric and magnetic fields are perpendicular to the energy flux direction.

The electromagnetic energy density $\varepsilon_{vort} = \left\{\epsilon_0 c^2 \mathbf{H} \cdot (\nabla \times \mathbf{E}) - \partial(c^2(\nabla \times \mathrm{H}) \cdot \mathbf{E}))\right\}$ is a spatial function, meaning that the spatial element of the medium would determine the energy. The electromagnetic energy flux $\mathbf{S} = \varepsilon_{\mathrm{dyn}}/c = \epsilon_0 c \nabla \cdot (\mathbf{H} \times \mathbf{E})$ is related to the rate at which the field energy moves around in space. The energy that flows through a small area ds per second is $\mathbf{S} \bullet \dot{n}$ds, where $\dot{n}$ is the unit vector perpendicular to ds. During the corpuscle electron's flight in the sample, the parameters, ω_t, k_t, r_t, $\mathcal{E}_t$, **D**, **B**, ϵ, and μ, of the sample medium would substitute the vacuum ones, resulting in a new electromagnetic energy density and rate of energy flux, $P_t = \nabla \cdot S = \nabla \cdot (E \times H)\ \partial_t \varepsilon_{EH} + \delta$. It should be noted that the $\mathcal{E}_{EH}$ is the energy density of the electromagnetic field in a vacuum and a function of the fields used to construct the energy flux vector. In general, it contains all the terms that can be expressed as a simple time derivative of the energy function. The δ terms are some residual contribution. If any change in energy flow **S** is balanced by the residual terms, $\delta \mathbf{P}_t = 0$, it means that there is no energy storage, but only energy flowing at every point in space. If transverse propagating fields are given by a unit vector $\boldsymbol{u}$ in the direction, then $\boldsymbol{E} \cdot \boldsymbol{u} = \boldsymbol{H} \cdot \boldsymbol{u} = 0$. This means we may construct directional fields such $K^{\pm} = \mathbf{E} \mp \mathbf{u} \times \mathbf{H}$ that it is a helical vector. Then

$$\mathbf{S} = \mathrm{u}\left(|\kappa^+|^2 - |\kappa^-|^2\right)/4$$

$$P_t = \nabla \cdot \mathbf{S} = 1/4 \partial_t (|\kappa^+|^2 + |\kappa^-|^2) + \boldsymbol{\delta}$$

If $P_t = 0$, that means the energy flux of the electromagnetic field of the photon in a corpuscle electron does not lose or gain energy with the vacuum. Then

$$(\partial_t - c\nabla \cdot \mathbf{u})\ |\kappa^+|^2 + (\partial_t + c\nabla \cdot \mathbf{u})\ |\kappa^-|^2 + = 4\boldsymbol{\delta}$$

This contains two counter-propagating wave components, each evolved by its own wave operator ($\partial_t + c\nabla \cdot$ **u**), which is a simple first-order wave equation for the intensity. The components of δ will act as source terms that drive and modify the otherwise simple linear wave propagation. The δ is closely related to the circumstances of the medium where the corpuscle electron is occupied.

This analysis shows that the confined electromagnetic energy density flux should exist as electromagnetic field energy flux hold in the space-time of the corpuscle contour. Therefore, the electron is an entity composed of an assembly of electromagnetic energy flux in confined space-time at the laboratory frame.

Based on this analysis, we said that the flying corpuscle electron traveling through the sample's medium should modify the rate of energy flux and rearrange the partitioning of energy flux in spatial distribution. This space-time process would encode the particular interaction between the corpuscle electron and the sample medium. We cannot directly measure the processes, but we may indirectly measure the rate variation of electromagnetic energy flux and momentum distribution in the spatial and temporal domain in TEM.

Velocity ds/dt contains space and time, and the speed of light c is the velocity metric based on the theory of relativity. The velocity of a flying electron in an electron microscope is related to the velocity of a dynamically circulating photon in its own corpuscle frame.

Understanding the relationship between these velocities is important for developing electron microscopy.

Based on Minkowski space-time, the invariant interval or space-time interval between P and Q events is defined as $\Delta s^2 = c^2\Delta t^2 - \Delta x^2$. This quantity Δs^2 can be positive or negative, so Δs might be imaginary. All inertial observers agree on the value of Δs^2. Therefore, we take this Δs as the "distance" between the two event points. For two infinitesimally separated motion events, the line element is

$$\Delta s^2 = c^2dt^2 - dx^2 - dy^2 - dz^2$$

and called word line. If $\Delta s^2 > 0$, it is called timelike, meaning it is possible to find inertial frames in which the two motion events occur in the same position and are purely separated by time. They can influence one another. If $\Delta s^2 < 0$, it is called spacelike separated, meaning the two motion events occur at the same time but are purely separated by space. They cannot influence one another. If $\Delta s^2 = 0$, it is called lightlike or null separated. The different points on the trajectory of a photon are lightlike separated. Note that $\Delta s^2 = 0$ does not imply that P and Q are the same event.

In Newtonian mechanics, we describe a particle by its position x(t), with its velocity being $u(\mathrm{t}) = dx/dt$. In relativity, this is unsatisfactory. In special relativity, space and time can be mixed together by Lorentz boosts, and we prefer not to single out time from space. For example, when we write the 4-vector X, we include both the time and space components, and Lorentz transformations are 4 × 4 matrices that act on X. In the definition of velocity, however, we are differentiating space with respect to time. First of all, we need something to replace time, which is why we define "proper length" as the length measured in its rest frame. Similarly, we can define "proper time." The proper time τ is defined such that

$$\tau = s/c$$

where τ is the time experienced by the particle, i.e., the time in the particle's rest frame.

The world line (or trajectory) of a particle can be parameterized using the proper time by $t(\tau)$ and $x(\tau)$. Infinitesimal changes are related by

$$d\tau = ds/c = (1/c)\left(c^2dt^2 - |dx|^2\right)^{1/2} = \left(1 - |u|^2/c^2\right)^{1/2} dt$$

Thus $dt/d\tau = \gamma_u$ with $\gamma_u = 1\ /(1 - |u|^2/c^2)^{1/2}$.

The total time experienced by the particle along a segment of its world line (or trajectory) is

$$T = \int d\tau = \int 1/\gamma_u dt$$

If the circulating photon moves with the speed of light, c, in the corpuscle frame (S′ frame), but this frame S′ (or corpuscle) moves with velocity v relative to the laboratory frame (microscope frame) S, what is its velocity u in the microscope frame S? The trajectory of the circulating photon in the corpuscle frame S' *is* $x' = u't'$, but in the laboratory frame S is

$$u = x/t = \gamma(x' + vt')/\gamma\left(t' + (v/c^2)x'\right) = u't' + vt'/\left(t' + (v/c^2)u't'\right) = u' + v/\left(1 + u'v/c^2\right)$$

This is the formula for the relativistic composition of velocities. The inverse transformation is found by swapping u and u' and swapping the sign of v i.e.

$$u' = (u - v)/\left(1 - uv/c^2\right)$$

It is worth noting that (1) if $u'v \ll c^2$, then the transformation reduces to the standard Galilean addition of velocities $u \approx u' + v$;(2) u is a monotonically increasing function of u′ for any constant v (with $|v| < c$); (3) This $u' = \pm c, u = u'$ holds any v, i.e., the speed of light c is constant in all frames of reference; (4) $|u'| < c$ *if* $|u| < c$. This means that we cannot reach the speed of light by the composition of velocities.

Using the geometry of space-time, one may clearly understand Minkowski space-time. Space-time has four dimensions, and each point can be represented by four real numbers. This can be seen when changing coordinate systems; instead of rotating the axes, we "squash" the time and space coordinate axes toward their diagonal, which is the axis of the speed of light c. The squash angle is related to the Doppler factor.

In the conventional space-time concept of an electron microscope, the transversal (horizontal) plane, which is perpendicular to the optic axis of the microscope, is a two-dimensional space, $\mathbf{r}$ (x, y), and may be called the hyperspace. The one dimension of time, ct, is the direction perpendicular to that plane. Each physical event is localized by a unique space-time point P(r, ct), while the evolution of an event is represented by a continuous curve called a trajectory or world line, which extends from its past to its future through its present (origin). The slope of the trajectory is

$$s = |\nabla_r(ct)| = c|\nabla_r t| = c/v$$

It is inversely proportional to the velocity (v) of the event. Thus, a vertical curve ($s \rightarrow \infty$) represents a stationary or static event, a straight line with $s > 1$ (or $v < c$) represents a (subluminally) uniformly moving object, a curved line (with subluminal $s > 1$ everywhere) represents a nonuniformly moving (accelerated or decelerated) object, and the $s = 1$ (or $v = c$), called a Dirac-like cone, represents light propagation in free space. The space {r,t} involves the Fourier direct space-time-independent variables **r** and t, which may be referred to as the direct space-time. The direct space-time also allows describing wave trajectories and scattering directions, but in the space of the inverse-Fourier, the space and time-independent variables k and ω logically may also be referred to as inverse space-time, which hosts the dispersion curves of the medium and provides a perspective that is obviously complementary to the parameters of the direct space-time.

In direct space-time, if a moment locus on the trajectory of a flying electron is at rest, then Δs=0, the circulating photon rotates with light speed on the light cone of the direct space-time, and its track path in two-dimensional space is a circle, which is the circumference of the Compton wavelength. But if the loci have velocity v, and $\Delta s < 1$, its track path is spacelike separated ellipses on which the tracks occur at the same time or simultaneity. They are purely separated by space and called the wavefront. The wavefront may be defined as the ensemble of all points which have the same phase angle, or in other words, it is simultaneity events. The wavelength, λ, is the spatial distance between the sequent simultaneity events or the points that have an identical phase. For a corpuscle electron, the wavefront of the circulating vorticity field of a photon is the ensemble of simultaneity's rotating velocity, which is ellipse contours. The wave vector, **k**, is a spatial vector on the ellipse wavefront describing the direction of the wave's propagation, and its magnitude is inversely proportional to the distance between corresponding points on the sequential wavefront, which is the wavelength. The wave vector **k** is a spacelike parameter, but it is related to the elapsed angular frequency, ω_k, of the vorticity field of the photon due to the corresponding phase related to the elapsed frequency, which is a timelike parameter. For angular frequency, ω, the $\Delta s^2 > 0$ is timelike, which means the two motion events occur in the same spatial position and are purely separated by time. Due to the events occurring at the same spatial point, the events can influence one another, which means the phase angle is different or the same. That is why the de Broglie wave can have interference and diffraction as optic waves do.

Due to Einstein and Planck's formula, $\varepsilon = mc^2 = h\omega$ and the de Broglie relation, $\mathbf{p} = h\mathbf{k} = h/\lambda$, these relations have distinctly unveiled the space-time characteristics: timelike is energy-involving mass, m, and angular frequency, ω, and space-like is momentum,

p, wave vector, **k**, and wavelength, λ. If using the speed of light, c, as the metric scale of space-time, then the time-like is ω/c, or Ɛ/h, and space-like is c**k** or **p**/h. For enunciating the space-time relation of $p_x/\hbar$ in the phase factor of a corpuscle electron, we may define the space-time position vector as x, whose time component $x^0 \equiv ct$. If we also define a space-time wave vector k, whose time component $\mathbf{k}^0 \equiv \omega/c\,(\mathbf{k}^\mu = (\omega/c;\mathbf{k}))$, is this ubiquitous phase factor may be written as a space-time dot product, $e^{i(-\omega t+kx)} = e^{i\mathbf{k}\bullet\mathbf{x}}$. Similarly, in an electron microscope, the wave function of a corpuscle electron (as previously mentioned) with definite momentum p and energy Ɛ moving in a vacuum is proportional to $e^{i(-Ɛt+px)/h}$. If we define a 4-momentum p with a time component, $\mathbf{p}^0 = \varepsilon/c\,(\mathbf{p}^\mu = (\varepsilon/c;\mathbf{p}))$, then this phase factor may also be written as a space-time dot product, $e^{i(-\varepsilon t+kx)h} = e^{ip\bullet\mathbf{x}/\mathbf{h}}$. The similarity between the two expressions already hints at the dual nature of particles and waves that is characteristic of quantum mechanics. When quantities like **p** • **x** are very large compared to ħ (many quanta), the effects of interference are numerically small, and it turns to "classical mechanics." Yet when $p\bullet x/h$ is of order unity or smaller, the corpuscle electron can display "wavy" behavior. The p • x can be expressed as $p\bullet x = mv^2(x/v) = 2\varepsilon_k\bullet t$ the product of time and the kinetic energy of circulating electromagnetic energy flux of the photon. It is worth noticing that the phase of the wave function actually implies the veiled electromagnetic energy flux.

3.4 Relationship between phase of wave function and internal dynamic process of a corpuscle electron

In an electron microscope, any trajectories of a flying corpuscle electron consist of many elementary spatial displacements, ds. If choosing length and time standards as $ds_0 = \hbar/mc$ and $\tau = ds_0/c = \hbar/mc^2$, while the mass conservation holds, the phase is proportional to the particle's mechanical action Š as $\varphi(r,t) = \check{S}(r,t)/h$. The space and timescale on the trajectory may be expressed as $\mathfrak{r} = r/ds_0$ and $\vartheta = t/\tau$. The phase value is integral on the interval $(\vartheta_1 - \vartheta_2)$, i.e., adding the rotation angular from t_1 to t_2. As previously discussed, the circulation of electromagnetic energy flux is a helix curve which may be expressed as

$$x = (\mathfrak{r}/2)\cos(2\varphi)\cos\beta\, y = (\mathfrak{r}/2)\sin(2\varphi)\, z = vt - (\mathfrak{r}/2)\cos(2\varphi)\sin\beta$$

The element displacement on the trajectory would be as

$$d\dot{\check{s}}^2 = \mathfrak{r}^2 d\varphi^2 + 2\mathfrak{r}\sin 2\varphi\sin\beta\, d\varphi\, vt + v^2t^2$$

If the $d\check{s} = ct$, then

$$c^2 = \mathfrak{r}^2\omega^2 + 2v\mathfrak{r}\omega\sin 2\omega\, t\sin\beta + v^2$$

That holds for flying corpuscle electrons, and the phase between t_1 and t_2 would be as

$$\varphi = \pm (c/\mathfrak{r}) \int \left(1 - (v/c)^2\right)^{1/2} dt$$

The ± signs indicate the right or left helicity. The action function of a free-flying corpuscle electron at the loci of the trajectory would be as

$$\check{S}(r,t) = h\varphi = -mc^2 \int \left(1 - (v/c)^2\right)^{1/2} dt$$

Therefore, we may have a specific geometric explanation of the corpuscle electron in terms of special relativity; the locus's element may be expressed as

$$(\mathfrak{r}d\varphi)^2 \equiv d\dot{\check{s}}^2 = c^2dt^2 - dz^2$$

That indicates the "space-time interval $d\dot{\check{s}} = \mathfrak{r}d\varphi$ in the corpuscle electron frame has the meaning of an arc length of the ultimate circumference, which is the immobile corpuscle's contour boundary in the plane of rotation of the photon. For a free-flying corpuscle electron in its own frame, this arc length is a definite unchanging number, which is treated as an invariant of special relativity. In comparison to standard relativity, this corpuscle electron has the time coordinate "coiled up" into the ultimate circumference, which is an analog of the Minkowski diagram. The Minkowski diagram acquires a fusion space-time form when the cylindrical helix is unrolled on a plane, but the difference in topological form is such that the "world line" is depicted by a point on the corpuscle border not by the electron itself. The line element is found to be proportional to a differential of the action function:

$$-mc\, d\dot{\check{s}} = \hbar d\varphi = -mc^2dt\left(1 - (v/c)^2\right)^{1/2} = d\check{S}$$

For free-flying electrons in an electron microscope, they can be regarded as either quantum or classical. If we think of the electron in the electron microscope as a quantum particle, then the action function would be as follows:

$$\begin{aligned}\hbar d\varphi &= \hbar(\partial\varphi/\partial t)dt + \hbar(\partial\varphi/\partial z)dz \rightarrow \\ &= \hbar\omega dt + \hbar k_n dx_n \\ &= -mc^2dt\left(1 - (v^2/c^2)\right)^{1/2} \\ &\cong -\varepsilon t + p_n dx_n\end{aligned}$$

Therefore, that automatically gives de Broglie's energy-frequency and momentum-wave vector relations as

$$\varepsilon = \left(mc^2 + 1/2\, mv^2\right) = \hbar|\omega| \text{ and } p_n = \hbar k_n$$

This free-flying quantum particle has an evident state function, usually called a wave function.

$$\psi = \psi_0 e^{i\varphi} = \psi_0 \exp\{i(1/\hbar)(p_n x_n - \varepsilon t)\} = \psi_0 \exp\{i(1/\hbar)(k_n x_n - \omega t)\}$$

From this analysis, it may be clear what the physical meaning of the phase of a wave function of a corpuscle electron is and what the de Broglie electron wave is.

It is well known from Fourier analysis that the Dirac δ-function can be expressed by an integral in the following form:

$$\delta(\varphi) = (1/\pi)\int_0^{\infty} \cos(s\varphi)ds$$

For an electron microscope, an electron is an energetic particle composed of confined electromagnetic energy flux that may be expressed as a δ-function, realized as the continual superposition of electromagnetic harmonic waves. That would not be possible to realize experimentally, but we can approximate the integral formula by the summation formula as follows:

$$\delta(\varphi) \approx (1/\pi)\Sigma_0^{\infty}\cos(s_i\varphi)$$

The electron-carried electromagnetic energy flux may be expressed as

$$\langle S\rangle_{av}\delta(\varphi)\delta(z-v_t t) = \epsilon_0 c\langle E^2\rangle_{av}\delta(\varphi)\delta(z-v_t t)$$

$<S>_{av}$ is the Poynting electromagnetic energy density flux. That indicates if the phase is zero, which is $(k_n x_n - \omega t' = 0$ at the corpuscle frame, the $\delta(\varphi) = 1$ so electromagnetic energy has a maximum at a locus of its trajectory in an electron microscope which is at $z_i = v_t t_i$. At electron microscope, the z-axis is usually visualized as the optic axis, but the velocity directions of the flying electron corresponding to the optical axis can be manipulated from 10^{-2} to 30°. Then, v_t may be expressed as a combination of parallel (longitudinal) and perpendicular (transversal) to the optical axis:

$$v_t = v_{\|} + v_{\perp} \quad v_{\|} = v_z \quad 1 = v_{\|}/v_t + v_{\perp}/v_t = \cos^2\alpha + \sin^2\alpha \quad \alpha \text{ is the divergent angle.}$$

Because the flying electron in an electron microscope travels around the optic axis, v_t is close to $v_{\|}$ and $v_{\perp}$ is very small in imaging mode. The path of the trajectory of the flying electron may be curving or have a larger divergent angle, causing the transversal velocity $v_{\perp}$ to become larger. The sample is located on the transversal plane, and the time for the electron traveling travel through the sample is dominated by the longitudinal velocity. Therefore, the time axis in an electron microscope may be chosen as parallel to the optical axis. The spatio-temporal contours of a flying electron on its trajectory may be expressed as

$$x = (\hbar/2m_0c)\gamma\cos\theta \quad y = (\hbar/2m_0c)\gamma\sin\theta \quad z = (v_t/\gamma)t$$

θ is the angle between the electromagnetic energy density flux direction and the z-axis (or optical axis). The x, y, and z are the coordinate values in the laboratory frame, but they correspond to the values in the corpuscle electron frame as

$$x' = x/\gamma \quad y' = y/\gamma \quad z' = z \quad t' = t\gamma$$

where t′ is the time in the internal frame of the corpuscle electron and the electromagnetic energy density flux circulates with a periodic time t′ as τ_0 which is the periodic rotation time of the electromagnetic energy flux at the rest condition of an electron. As a locus on a trajectory displaced by Δs with a time interval Δt in the laboratory frame, simultaneously in the internal frame of the corpuscle electron, the electromagnetic energy flux would have a slip/helical spiral path with $\Delta z'$ and $\omega' t'$ in space and time in the internal frame of the corpuscle electron. The instantaneous position of the corpuscle electron in the laboratory frame (or microscope frame) has to be spread out over the locus's coordinates and results in a finite non-zero location (scale around 10^{-13} m) that limits the accuracy of measuring the location in the microscope frame.

It should be noted that the transversal contours of the flying corpuscle electron are directly related to the slanting angle to the velocity direction in space, and longitudinal velocity dominates time for slipping motion. Based on the Poynting theorem, the longitudinal direction is the electromagnetic energy flux direction, but the electromagnetic fields are defined in the transversal plane, which is the sample plane in TEM. Therefore, the interaction between the transmitting electron and the sample at TEM occurs in the transversal space, but the interacted information in the transversal space is encoded in the transmitted electromagnetic energy flux.

3.4.1 Electromagnetic fields of a corpuscle electron at space and time

As mentioned before, the wave function of an electron can be expressed as an electromagnetic field, and the Maxwell equation may correspond to the Schrödinger equation. So, we use the wave function as an electromagnetic field to discuss the interaction between the electron and the sample. The de Broglie wave may be seen as the Doppler pulsating wave of the rotation of the circulating electromagnetic vorticity field of a photon. The electromagnetic vorticity field of a photon may be visualized as harmonics, and the rotated harmonics build the electromagnetic pulsating wave fields of the corpuscle electron with symmetrically distributed wave vectors and frequencies while it is fly-free. The spatio-temporal pulsating field dynamics may be expressed in a Minkowski space-time diagram as shown in Figure 3.1. It shows that the three dimensions of a corpuscle electron in the laboratory frame may actually be seen as two dimensions, with the third one as the time axis scaled by the speed of light, c.

The spatial contour of the corpuscle electron at rest is determined by the dynamic circulating photon. The frequency of the circulating photon is about 10^{20} Hz (or temporal periodicity is about 10^{-20} s or $\tau_0 = 10^{-20}$ s) and its rotating loop radius is about 10^{-13} m or 10^{-4} nm. These data indicate that a corpuscle electron is a spatio-temporal pulse, which contains confined dynamic electromagnetic energy circling flux with super-high frequency (10^{20} Hz) and 10^{-4} nm spatial contour.

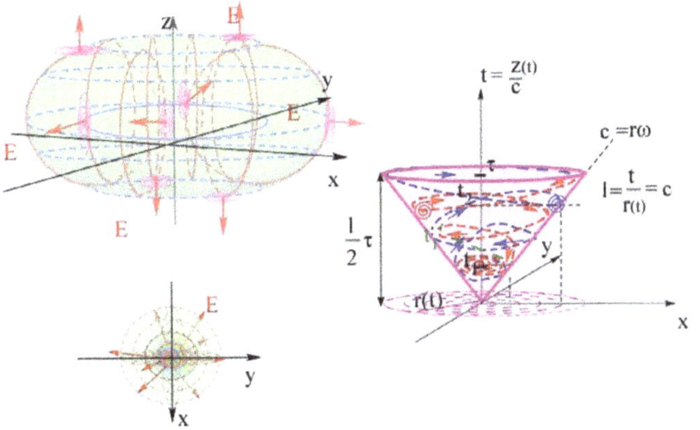

Figure 3.1: In corpuscle electron frame, the dynamic motion of the electromagnetic vorticity field of a photon is depicted. In the left column, the corpuscle electron is at rest, while in the right column, it shows the electromagnetic energy flux with light-speed circulation on the optic cone with time periodicity τ in the Minkowski space-time diagram. The blue- and red-dotted lines indicate the dynamic movement at half-time periodicity.

3.4.2 Electromagnetic field characters of a flying corpuscle electron

For an electron microscope, the corpuscle electron flies with high velocity (usually from 0.5c to 0.9c). In the electron microscope frame, the measured physical parameters of the electromagnetic fields of a flying corpuscle electron originate from the dynamic electromagnetic fields of a circulating photon.

Figure 3.2 demonstrates a dynamic motion along the temporal axis, which is perpendicular to the two-dimensional spatial plane. The fields of vorticity photon have a circular distribution of the wave vectors. If the corpuscle electron flies with $0 < v/c < 1$, the laboratory observer will see a different distribution of the wave vectors and frequencies of the dynamic circulating photon as shown in Figure 3.3.

From Figure 3.3, we may understand that the flying electron is a rotating bullet with dynamic momentum $\mathbf{p} = m\mathbf{v} = \gamma\, m_0\mathbf{v}$. Then we may deduce

$$\mathbf{p} = \gamma\, m_0 c^2 (\mathbf{v}/c^2) = \varepsilon\, (\mathbf{v}/c^2)$$

If the energy is electromagnetic wave energy $\varepsilon_{em} = h\omega$ then $\mathbf{p} = h\omega(\mathbf{v}/c^2)$. Based on the de Broglie relation $\mathbf{v} \cdot \mathbf{v}_{ph} = c^2$ we may have $\mathbf{p} = h\omega/\mathbf{v}_{ph} = h\mathbf{k}$ that indicate that the flying corpuscle electron has a symmetric elliptical momentum distribution around the longer axis of the elliptical circle. The symmetric swirling elliptical momenta make the flying

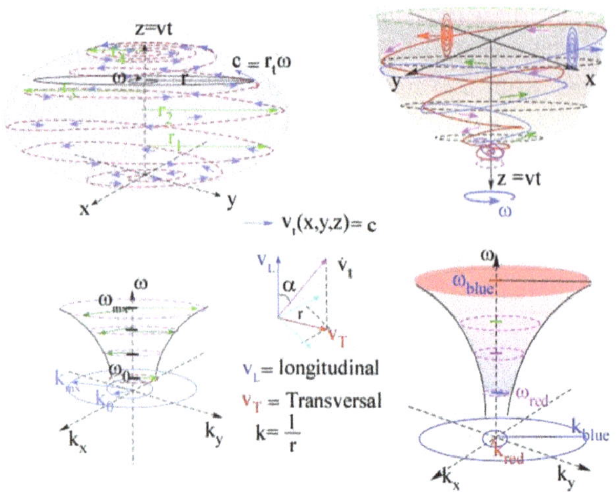

Figure 3.2: The circulating frequencies and wave vectors of the circulating electromagnetic field of the dynamic photon at the corpuscle electron frame are related to each other based on the flying velocity. The left column shows the two perpendicular rotation frequencies corresponding to the rotation frequency ω_0 and the frequency of the circulating photon $\omega_{mx} = 2\omega_0$ and wave vectors. The right column shows that the flying velocity induces the modification of these frequencies and wave vectors. The flying velocity along the z-axis or time axis induces the extended frequency and circling wave vector distribution.

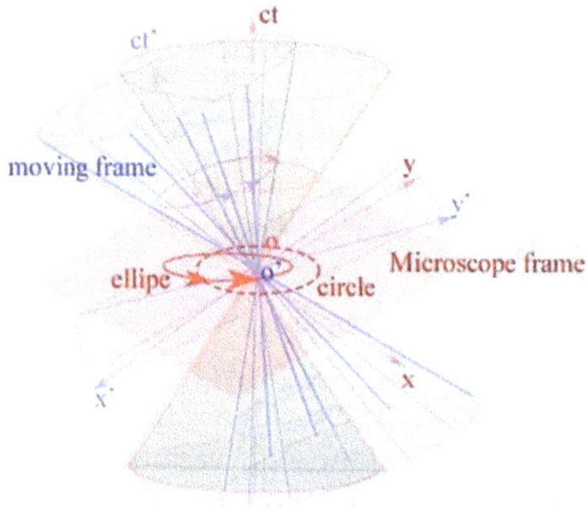

Figure 3.3: The Minkowski space-time diagram shows the wave vector distribution in the microscope frame and the flying corpuscle electron frame. In the electron microscope frame, it is a circular wavefront, which represents the same dynamic states or wave vectors simultaneously at rest for the corpuscle electron. It is an elliptical circle for the flying corpuscle electron. The longer axis direction of the elliptical circle corresponds to the flying direction.

direction of the corpuscle electron invariant. If the symmetric swirling is broken, the flying direction of the corpuscle electron would change its moving direction as shown in Figure 3.4.

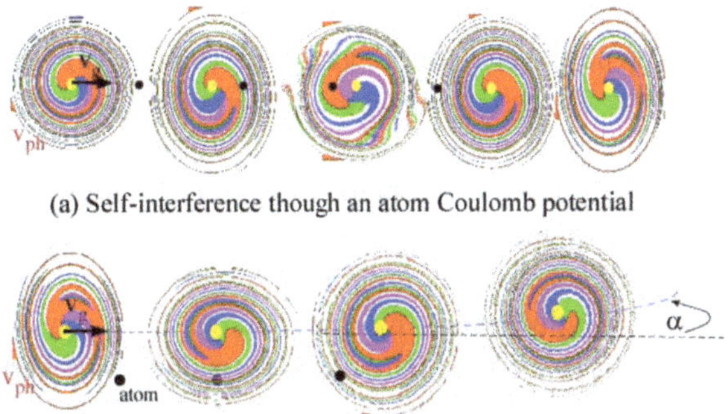

Figure 3.4: A corpuscle electron passes through or bypasses an atom's Coulomb potential: (a) symmetric pass-through as self-interference and (b) bypasses the atom as scattering in the laboratory frame.

It is clear that the internal swirling electromagnetic energy flux dynamics dominate the corpuscle electron motion in the laboratory frame. This dynamic electromagnetic energy flux flows symmetrically around its center of energy balance when freely flying in a vacuum, except when the balance is broken by an external field or medium. For electron microscopy, the electron beam is the corpuscle electron flowing current, and at any given moment, it is impossible for two corpuscles to be in the same spatial location. However, at the same spatial location, there can be many corpuscle electrons with different elapsed times in the laboratory frame, where the spatial scale is larger than 10^{-13} m. Due to the electron spin's magnetic interaction between electrons, the separation intervals may vary. From my view, the electric field of a corpuscle electron may be a swirling electric field with a super-high frequency as a pulsating field, which appears as a stationary electric field to a laboratory observer. The interaction between light and electrons is possible.

In an electron microscope, the trajectory of a corpuscle electron may be solved by using the energy flow conservation equation, $\varepsilon = mv^2/2 + \varepsilon_{res} = \text{constant}$ and the element displacement of the trajectory can be expressed using the action function as mentioned before:

$$\check{S}(r,t) = h\varphi = -mc^2 \int \left(1 - (v/c)^2\right)^{1/2} dt$$

Then the Hamilton-Jacobi equation is

$$\partial_t S + (\partial_n \check{S} \partial_n \check{S})/2m + \varepsilon_{res} = 0$$

These solutions would give the path of the trajectory of different flying electrons.

As discussed in Chapter 1, the energy of a corpuscle electron contains two parts: the rest energy $m_0c^2 = h\omega_0$ and the kinetic energy, $\varepsilon_{kin} = \gamma m_0 c^2 - m_0 c^2 = m_0 c^2(\gamma - 1)$. The kinetic energy comes from the corpuscle electron moving with velocity v_t in an electron microscope. The kinetic energy is classical mechanical energy, and the total energy of a corpuscle moving with velocity v_t is $\mathcal{E} = \mathcal{E}_0 + \mathcal{E}_{kin.}$ However, the law of energy conservation tells us that $\varepsilon^2 = (\varepsilon_{kin})^2 + (m_0c^2)^2$. As stated in Chapter 1, the Poynting vector is the momentum of the electromagnetic field flux, and the circulation of the Poynting vectors builds the confined electromagnetic energy flux in the contours of a corpuscle such as an electron. If we designate this electromagnetic energy as the pulsating energy of the corpuscle electron, which is $\varepsilon_{puls} = (\gamma m_0 v)c = mvc$ and $\varepsilon^2 = (\varepsilon_{puls})^2 + (m_0c^2)^2$. From $\varepsilon = \varepsilon_0 + \varepsilon_{kin}$ and $\varepsilon^2 = (\varepsilon_{puls})^2 + (m_0c^2)^2$ we can deduce the following relation:

$$v = c\left(1 - \left[(\varepsilon_0)^2/\varepsilon^2\right]\right)^{1/2} = c\left(1 - \left[1/(1 - \varepsilon_{kin}/\varepsilon_0)^2\right]\right)^{1/2} = c(\varepsilon_{puls}/\varepsilon) = c\left[\varepsilon_{puls}/(\varepsilon_0 + \varepsilon_{kin})\right]$$

That relationship indicates that if the corpuscle has a state of $\mathcal{E}_0 = \mathcal{E}$, then $v = 0$. The velocity v dominated $\mathcal{E}_{puls}$ instead of $\mathcal{E}_{kin}$. In other words, the momentum of a flying corpuscle electron is mainly a pulsating momentum instead of linear inertia momentum.

As the de Broglie relation indicated $\mathbf{v}_{ph}\mathbf{v}_g = c^2$ the group velocity, $\mathbf{v}_g$, is the moving velocity, v, of a corpuscle electron $\mathbf{v} = \mathbf{v}_g = c\left[\varepsilon_{puls}/\varepsilon\right]$.

Therefore, due to $v = ({v_x}^2 + {v_y}^2 + {v_z}^2)^{1/2}$ the electron microscope, the corpuscle electron flying along the optic axis or z-axis may be expressed as $v = v_z \exp(i\theta)$, in which $\theta = 0$, $v = v_z = v_{||}$ (longitudinal), and $\theta = \pi/2, v = v_\perp$ (transversal). If $\theta = 0$, then the flying corpuscle electron with velocity $v = v_z = v_{||}$ would be a pulsating energetic corpuscle with frequency $\omega_{puls} = \omega(1 - (\omega_0/\omega))^{1/2}$ in which $\omega = \omega_0[(1 + v_z/c)/(1 - v_z/c)]^{1/2}$ is a super high-frequency pulse in the z-axis or optical axis of the microscope. For the transversal velocity component $v = v_\perp$ the pulse frequency is $\omega = \omega_0[1 - (v_\perp/c)^2]$ lower (or Doppler red shift). Therefore, in the direction of the flying corpuscle electron, v_z has a very short de Broglie wavelength for TEM, for example, at $100\text{kV}\lambda_{db} = 0.0038$ nm kV; at $300\text{kV}\lambda_{db} = 0.0022$ nm kV. However, in the transversal direction, the electromagnetic energy wave may have a long wavelength depending on the flying velocity v_t slanting angle to the optical axis. For example, at 100 kV and a slanting angle of 10^{-3} radian, the transversal frequency is about $0.7\omega_{0,}$ and at 300 kV, this frequency is about $0.3\omega_0$. The frequency ω_0 is related to the internal rotation frequency of the stationary corpuscle electron at its internal frame, and for interaction between the harmonic circulating fields, the lower frequencies make energy exchange easier by frequency modification and circulation phase match, which provides the spatio-temporal condition for energy conservation and circulating phase match.

For a transmission electron microscope, the flying corpuscle electron with velocity v_t near the optical axis should be understood as an electromagnetic energetic pulse with super high frequencies or energy along the parallel to the optical axis. Simultaneously, its transversal electromagnetic energy field has a variable lower frequency corresponding to its slanting angle to the optical axis. The flying corpuscle electron observed along its flying direction at the microscope frame is an electromagnetic energetic quantum particle with a space scale of about 10^{-3} nm and an elapsed duration of $<10^{-20}$s. Its transversal electromagnetic fields have lower energy and frequencies at the spatial scale of about 10^{-3} to a few nanometers. This physical nature of the electron would dominate the phenomena observed in TEM.

3.4.3 The interaction between the corpuscle electron and sample medium

The phase of the de Broglie electron wave $\varphi = 1/\hbar(\varepsilon_v t - px)$ can be expressed as

$$\varphi = 1/\hbar\left(\varepsilon_v t - px\right) = (1/\hbar)\left(\varepsilon_0\left(1-(v/c)^2\right)^{-1/2}\right)\left(t-(v/c^2)vt\right)$$
$$= \left(1-(v/c)^2\right)^{1/2}(\varepsilon_0/\hbar)t = (\varepsilon_0/\hbar)t'$$

and

$$t' = \left(1-(v/c)^2\right)^{1/2} t \text{ and } \varepsilon_v = \varepsilon_0\left(1-(v/c)^2\right)^{-1/2}$$

That clearly shows that the phase of the de Broglie electron wave is closely related to the electromagnetic energy of the circulating photon and Planck constant and dynamic time at the corpuscle electron frame. The time observed at the electron microscope is not the same as the time at the corpuscle electron frame. The ratio of time t′ at the corpuscle electron frame and the time t at the laboratory frame is the $\gamma = (1-(v/c)^2)^{-1/2}$. The electromagnetic energy carried by an electron should be attributed to circular rotation originating from the rest state, which is the energy core of an electron, and hyperbolic rotation, which comes from the kinetic motion of an electron. The electromagnetic energy core of a corpuscle electron is hard to change, but the kinetic motions can be changed by the variation of direction and magnitude of the kinetic velocity, which is the way to play for electron microscopy.

For understanding, intuitively, the physical entity of a corpuscle electron in a microscope, we may use Minkowski geometry to illustrate the relationship between the internal de Broglie phase waves and electromagnetic energy flux dynamic motions for an elapsed locus of a trajectory of the corpuscle electron with transient velocity v as shown in Figure 3.5.

The core energy of a corpuscle electron is above $\varepsilon_0 = m_0c^2 = 0.5109989$ MeV, but the kinetic energy may be such $\varepsilon_v = (v_t/c)^2\varepsilon_t$ that if the kinetic motion vanishes,

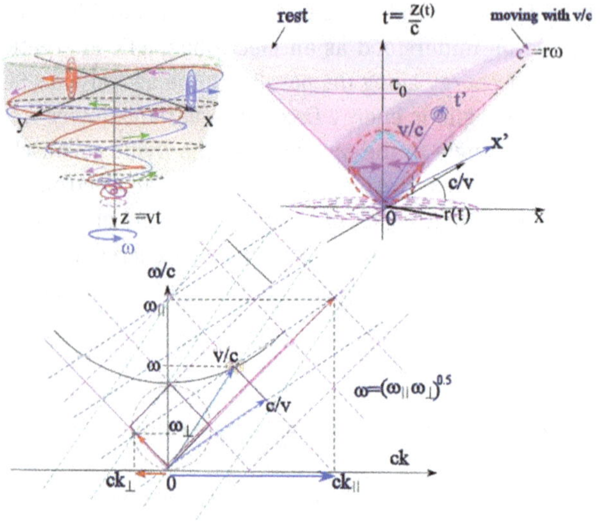

Figure 3.5: Minkowski geometry shows a corpuscle electron at a locus of trajectory with transient velocity v. The $ck_{\parallel}$ and $ck_{\perp}$ are the longitudinal (Doppler blue shift) and transversal (Doppler red shift) wave vectors.

there is only the core energy of the corpuscle electron. This implies that the captured electron transforms the core energy to the detector, and the kinetic energy can only be detected by measuring the velocity of a corpuscle electron, which is used as a velocity spectrum (for example, the electron energy spectrum).

The fusion of circulation and rotational motions of the electromagnetic field induces the merging of spatio-temporal processes that result in the particle-wave duality of an electron. The split of the spatio-temporal process should be the target of electron microscopy because an electron microscope only magnifies the spatial dimension but not the temporal scale directly. Of course, the spatial scale magnified also shows the effect of the timescale indirectly by $t = s/v$; while the spatial distance s is expanded, the timescale is increased too. Therefore, understanding the spatial and temporal processes would be important for understanding the interaction between the electron and the sample.

3.4.3.1 Relationship between circulation, kinetic motion, and time

We already mentioned that the phase ωt may be analyzed as

$$\varphi = \omega t = \left(1-(v/c)^2\right)^{-1/2}(\varepsilon_0/\hbar)t = \left(1-(v/c)^2\right)^{-1/2}\omega_0 t = 2\pi t'/\tau_0$$

where τ_0 is the time period of the rotation process of the circulating photon at rest in its internal frame. The time t is the elapsed time on the trajectory of the flying corpuscle electron in the laboratory frame. It is better to express this as Δt_i, which shows $\Delta\varphi_i = 2\pi\Delta t'_i/\Delta\tau_0$. Due to $\Delta\tau_0(1-(v_t/c)^2)^{-1/2} = \Delta t'_i$ then $\Delta\varphi_i = 2\pi\Delta t'_i/\Delta\tau_0 = \gamma = (1-(v_t/c)^2)^{-1/2}$ that the phase of the wave function is related to the transient velocity of the corpuscle electron. In other words, the transient velocity of a flying corpuscle electron is related to the corresponding rotating angle of the circulating photon in its internal frame. This phase difference is the effect of relativity.

de Broglie matter wavelength, λ_{db}, may relate to the Compton wavelength. This indicates $\lambda_{db} = \lambda_C/(v/c)$. that if the electron velocity is smaller than the speed of light, c, and the de Broglie wavelength is longer than the Compton wavelength of an electron. The Compton wavelength reveals the internal rotation of the circulating photon, but the de Broglie wavelength expresses the additional kinetic motion of the circulating photon, which modifies the Compton wavelength. Therefore, if $v_{||} \approx v_t$ and $v_\perp \ll v_{||}$, then the wavelength along $v_{||}$, or longitudinal velocity is shortest, but the transversal one is longest, as discussed in the previous section and shown in Figure 3.6.

The additional kinetic motion may express as spatial orientation variation of the Poynting vectors k(t′) with carrying different energy flux that shows as the spatial phase component $\mathbf{k}(t')\mathbf{r}'(t')/\hbar$. The spatial elliptical distribution of Poynting vectors may lead to the electromagnetic energy density flux re-splitting in different spatial orientations $\varepsilon = \sum_i \varepsilon_i(k_i)$. Due to the circular and elliptical kinetic motion of the vorticity field of a photon, the spatial and temporal symmetry of the electromagnetic field is held, and the flying direction is steered by the electromagnetic momentum density flux $\mathbf{S} = 1/c\ (\mathbf{E} \times \mathbf{B})$. If the electromagnetic momentum flux is symmetrically disturbed, the flying electron may not change traveling direction of the corpuscle electron. However, the asymmetrically disturbing electromagnetic field would modify the traveling direction of the corpuscle due to the de Broglie relationship:

$$\mathbf{v}_{ph} \cdot \mathbf{v}_g = c^2 \text{ and } \mathbf{v}_g = \nabla_k \omega(k(t)) = \nabla_k \varepsilon_k(t)/h$$

As shown in Figure 3.4, the gradient direction of the energy flux or potential will rectify the moving direction of the corpuscle electron.

3.4.3.2 Electromagnetic characteristics of sample

It is well known that any matter is an electromagnetic medium because it consists of atoms that construct different spatial symmetric configurations, such as crystals or amorphous structures. The spatial periodicity (long or short range) is an essential characteristic. The intrinsic closed-packing planes of the atoms are stacked consecutively, building up the crystal lattice. The primitive lattice unit cells have the largest separation between the packing atoms, and the primitive reciprocal unit cell is the first Brillouin zone, which exhibits the density distribution of the electromagnetic field of the atoms. This distribution dominates the physical parameters, for example,

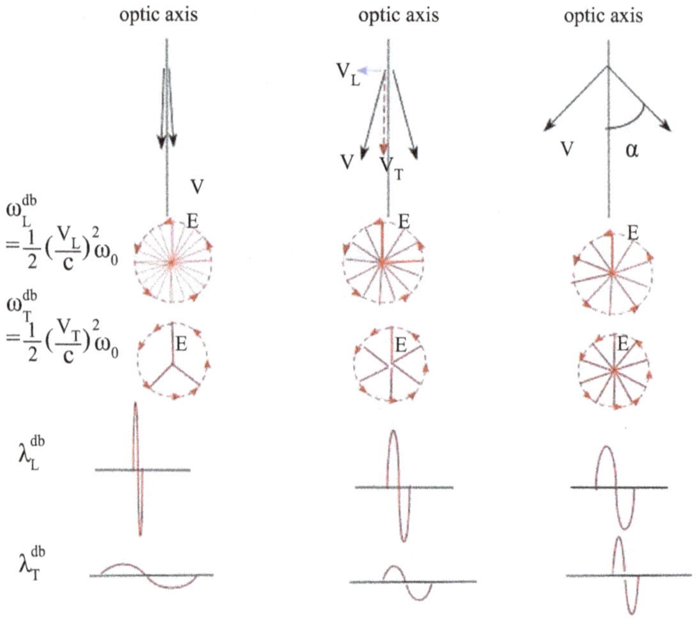

Figure 3.6: The longitudinal and transversal components of the velocity of the corpuscle electron steer the angular frequency and de Broglie wavelengths.

permittivity ε and μ, of the sample. As is well known, the permittivity is related to the speed of light, as $c = (\varepsilon_0\mu_0)^{-1/2}$ in a vacuum, but $v = c/\check{n} = (\varepsilon\mu/\varepsilon_0\mu_0)^{-1/2}$ in a medium. Therefore, while the corpuscle electron is viewed as a confined energetic electromagnetic energy pulse, using an electron to detect the physical properties of matter is possible, except to exhibit the periodicity of the assembled atoms.

It is possible to use a periodic function of a spatio-temporal phase variable to express the permittivity of a crystal:

$$\varepsilon = \varepsilon_0 f(\Theta) \text{ where } \Theta = \{\aleph Z - \dot{\omega} t\}$$

where ϰ being the spatial modulation frequency and $\dot{\omega}$ being the temporal modulation frequency. These frequencies may be related to the dynamic motion of the component atoms in a sample. Due to ε_0 being the vacuum permittivity, the permittivity of a solid ε may expand to an assembly of the infinite electromagnetic waves of resonators of periodic atomic planes located parallel to the optical axis of the microscope as $f(\varkappa, \dot{\omega}) = \sum n_i \exp(\Theta n_i)$, ϰ is the spatial frequency of the periodic atomic planes, and $\dot{\omega}$ is the resonator frequency of the periodic atomic planes. The physical parameters of the solid are dominated by the variables ϰ and $\dot{\omega}$. It is known that the transmitting velocity v of an electromagnetic wave in a medium is dominated by the index of refraction $\check{n} = c/v_t$. This means the corpuscle electron transmitting velocity v_t in a medium would be modified by the index of refraction of the medium, which would rectify the wave transmitting direction $v = \omega/k'$, resulting in the diffraction of the flying corpuscle electron by the solid.

As mentioned in the previous section, in the space-time of a corpuscle electron frame, the two-dimensional electromagnetic vorticity field of a photon has the electromagnetic energy density flux as the Poynting vector $\mathbf{S} = 1/c\ (\mathbf{E} \times \mathbf{B})$. This energy density flux is always circling at the speed of light c on its light cone in Minkowski geometry. While the electron flies with velocity v along the optic axis in an electron microscope, the cone of light is skewed at an angle (μ), called the velocity parameter or hyperbolic angle, related to v/c as tangh (μ) = v/c and this creates the special relativity's elliptical wavefront, which is a slicing at the angle parallel with the plane of simultaneity of the photon in the corpuscle frame. The electromagnetic energy density flux $\mathbf{S} = \varepsilon_{dyn}/c = \epsilon_0 c \nabla \cdot (\mathbf{H} \times \mathbf{E})$ would be circulated and displaced along the elliptical wavefront in space with time. The instantaneous directions construct a cone-like wavefront along the flying velocity of the corpuscle electron, and the cone angle is the hyperbolic angle μ. The instantaneous velocity of the flying electron is the projection of the speed of light in the direction of the flying electron. If the instantaneous Poynting electromagnetic energy density flux could be expressed as an assembly of many plane waves with wave vector k_i and its instantaneous propagating direction k_i angle θ_i with the optical axis z at in the electron microscope frame. Then in the corpuscle electron frame the angle is θ_i' and we may have the following relation between the electron microscope frame and the internal frame of the corpuscle electron:

$$\cos\theta_i = [\cos\theta_i' + (v/c)]/[1 + (v/c)\cos\theta_i']$$

For an electron microscope frame at a given acceleration voltage, for example, 300 kV, the velocity of the electron along the optical axis z is constant; so (v/c) is constant. The instantaneous Poynting energy flux varying in directions corresponding to the z-axis in the internal frame would act as the electromagnetic energy flux in the electron microscope frame. This means the periodicity of the Poynting energy flux propagating with the elliptical wavefront in the corpuscle electron frame is the origin of the de Broglie electron wave. The de Broglie electron wave consists of dynamic electromagnetic waves with a helical velocity distribution, where the velocity along the optical axis (or longitudinal velocity) is the largest, but the velocity perpendicular to the optical axis (or transversal velocity) is the slowest. This is due to the Doppler effect as shown in Figure 3.5.

In a vacuum, at a corpuscle electron internal frame, the electromagnetic energy density flux propagates by oscillation between the electric and magnetic fields due to Maxwell's laws, $\nabla \times B = \partial_t E$; $\nabla \times E = \partial_t B$. For a full description of a magnetic field in Minkowski space-time, the magnetic field has a potential-vortex nature. The magnetic fields have a vortex vector field and a potential displacing field with vorticity. Therefore, at any point of the wavefront, there are two components of the electromagnetic field: transversal and longitudinal. Each of these components is almost planar, which means the wavefront of each wave coincides with a plane that is perpendicular to the direction of transmission. If the transverse electromagnetic wave distributes along the x-axis and the longitudinal one moves along the y-axis, within one wave spatial period, four consecutive stages can be distinguished as follows: (a) generating the vortex magnetic field H at the point M (x, y) during the time period, $0 \le t \le 1/4T_0$; (b) generating the vortex electric field E at the point M (x_1, y_1) during the time, $T_0/4 \le t \le 1/2\ T_0$; (c) generating the scale magnetic field (H^*) at M_2 (x_2, y_2) during the time, $T_0/2 \le t \le 3/4T_0$, and (d) generating the potential displacing electric field $E_\rightarrow$ at the point M_3 (x_3, y_3) during the time, $3/4T_0/ \le t \le T_0$. This process actually occurs in the x-y plane because both components have the same velocity, and the line where these points lie is a bisector of a right angle. The distance between adjacent points should be taken as equal to a quarter of a wavelength, λ/4, with each next stage occurring with a lag of a quarter of a wave period. We can express this process as the electric and magnetic fields exchanging with each other:

$$H\ (x,\ t) = H_z(x)\ \exp\ (i\omega t)$$

$$E_o(x_1,\ t_1) = E_{oy}(x_1)\ \exp(i\omega t_1)\ x_1 = x + (1/4)\lambda t_1 = t + (1/4)T_0$$

$$H^*(y_2,\ t_2) = H^*(y_2)\ \exp(i\omega t_2)\ t_2 = t + (1/2)T_0$$

$$E_\rightarrow(y_3,\ t_3) = E_\rightarrow(y_3)\ \exp(i\omega t_3)\ t_3 = t + (3/4)T_0$$

The transverse electromagnetic wave vector $k_\perp = \omega(\varepsilon_0\varepsilon'\mu_0\mu')^{1/2}$ and magnetic field are $H(r,t) = H_z{}^0 exp\, i(\omega t - k_\perp x) = H_z{}^0 exp\, i(\omega t - k_\perp x^0 \bullet r)$. Therefore, characteristics of an electromagnetic wave has four characteristics the transverse component of a wave, determined by the vortex vectors E_0 and H, and the longitudinal component, characterized by the potential vector $E_\rightarrow$ and H^* (E_o is the vortex field and $E_\rightarrow$ is the translating field). The velocities of transverse and longitudinal electromagnetic waves are the same, which imply that both components of an electromagnetic process are indissolubly linked, making it impossible to consider them separately in a general case. Transverse electromagnetic waves, which propagate in the physical environment, generate longitudinal waves at each point, and vice versa. The wave vector of a transverse electromagnetic wave is $k_\perp = \omega(\varepsilon_0\varepsilon'\mu_0\mu')^{1/2}$ and the wave vector of a longitudinal electromagnetic wave is $k_{||} = \omega_{||}(\varepsilon_0\varepsilon'\mu_0\mu')^{1/2}$. Both components of an electromagnetic wave transport energy, which may be expressed as

$$\mathbf{p} = \mathbf{p}_\perp + \mathbf{p}_{||} = E_o \times H + E_\rightarrow H^*$$

The direction of the resulting vector (Poynting vector) coincides with the radius vector drawn from the origin center of the corpuscle electron to the point where the field is defined. The function of an electromagnetic energy density looks like this:

$$\begin{aligned}\varepsilon &= 1/2(\mathbf{E}_0 \bullet \mathbf{D}_0 + \mathrm{HB} + \mathrm{H^*B^*} + \mathrm{E}_{\rightarrow} \bullet \mathbf{D}_{\rightarrow}) \\ &= (\varepsilon\varepsilon_0/2)\left[(\mathrm{E}_{oy}{}^0)^2 + (\mathrm{E}_{\rightarrow}{}^0)^2\right] \\ &= (\mu\mu_0/2)\ \left[(\mathrm{H}_z{}^0)^2 + (\mathrm{H^*0})^2\right]\end{aligned}$$

The first two terms characterize the energy of a transverse wave, and the last two terms describe a longitudinal wave. The transverse electric energy density is equal to the longitudinal energy density. However, both energy densities depend on the medium's characteristic parameters $\varepsilon\varepsilon_0$ and $\mu\mu_0$, which tell us that the electron in an electron microscope lives in a vacuum having ε_0 and μ_0. When the electron passes through the sample, which may be a metal, dielectric, or semiconductor, the energy density distribution of the electromagnetic vorticity field would be determined by the spatial and temporal characteristics of the sample medium, which are manipulated by the longitudinal and transversal permittivity $\varepsilon_{||}\mu_{||}$ and $\varepsilon_{\perp}\mu_{\perp}$.

The moving corpuscle electron consists of the kinetic helix vortices fields of electromagnetic waves with velocity v_z in the z direction (optical axis), and the electric field distributions in the de Broglie wavelength have longitudinal and transversal directions. The rotating vortices field energy flux consists of the dynamic circulating vorticity field with angular frequency $\omega_{zit} = ck_1$ and vortex rotation with angular frequency $1/2\omega_{zit} = ck_2$; the photon's electromagnetic energy $\mathcal{E}_{em}$, which is the energy flux contained in a corpuscle electron, flows in a $c\mathbf{k_0}$ direction, synthesized by the rotation and displaced kinetic motion. The velocity of the corpuscle electron is the projection of the electromagnetic energy flux $c\mathbf{k}_0$, as shown in Figure 3.7. It clearly shows that the electric and magnetic fields of the rotating vortices field of the photon are always perpendicular to that if transmission direction.

While the sample and corpuscle electron overlapped each other, the longitudinal and transversal components of the velocity of the corpuscle electron will play different roles. The longitudinal component determines the overlapping temporal scale, but the transversal component controls the spatial interaction process. The longitudinal velocity $v_{||}$ of a corpuscle electron is related to the blue shift of the Doppler effect, which shows the de Broglie's shortest wavelength (usually called the wavelength of a flying electron), and the transversal velocity $v_{\perp}$ of the corpuscle electron is dominated by the redshift of the Doppler effect, which has the longest wavelength. Steering the angle between the velocity and optical axis will modify the de Broglie wavelength in the longitudinal and transversal directions. The highest frequency will exhibit the energetic particle character of the corpuscle electron due to the difficulty in matching the phase of the rotation of the circulating photon, but low frequency or long wavelength makes the phase match possible. That is why electron micrographs always show two spatial dimensions, which are perpendicular to the optical axis, as a feature of a sample.

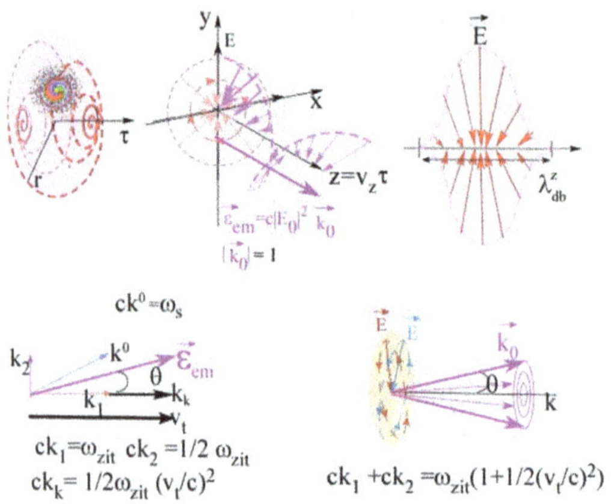

Figure 3.7: The moving corpuscle consists of the dynamic helix vorticity and radiated electric fields along the de Broglie wavelength in the direction of motion. The relationship between the S(2) × S(2) rotation of the electromagnetic vorticity field of the photon and the electromagnetic energy flux ck_0 is projected in the direction to determine the velocity of the corpuscle.

3.5 Traversing process of a corpuscle electron over a sample

3.5.1 A corpuscle electron traversing over a nuclei of an atom

de Broglie's relationship, $\mathbf{v}_g \cdot \mathbf{v}_{ph} = c^2$, shows us that the corpuscle electron's velocity in the laboratory frame is tightly related to the phase velocity of the internal oscillation of the confined electromagnetic energy in the internal frame of a corpuscle electron. Planck's constant, h, may be used as a ruler to measure the temporal energy density, $h = \varepsilon_t/\omega(\tau) = \varepsilon_t \delta\tau$, where t is the moment in the laboratory frame and τ is the corresponding time in the corpuscle frame. In this spatial dimension, Maxwell's equations of the electromagnetic fields dominate the electromagnetic interactions, and the super high frequency of the electromagnetic wave field for Zitterbewegung vorticity field is around ZHz $\left(\approx 10^{18} - 10^{-22}\text{Hz}\right)$. The timescale of about an attosecond ($\approx 10^{-18}$ s). The electric field envelope function $E_0(x, y, z, t)$ of an electron may be modified by the variation of the electromagnetic wave momentum flux vector (Poynting vector), triggering a change in the corpuscle electron's propagation direction in the laboratory frame. As mentioned in Chapter 1, in the rest state, the circulating angular frequency is $\omega_0 = d\theta/dt = m_0c^2/h$, the tangent linear velocity is $L_0\omega_0 = c$. When it is moving with velocity v_t along a spatial direction in the laboratory frame, there are

$$v_t/c = r\omega/r_0\omega_0 = (\omega/\omega_0)(r/r_0) = (h\omega/h\omega_0)(r/r_0) = (m/m_0)(r/r_0)$$
$$= (m/m_0)[r/(h/m_0c)] = r(mc/h) = rp/h = kr\ (\text{or}\ k \cdot \mathbf{r})$$

This indicates that the ratio of the speed of light and the velocity of the corpuscle electron in the laboratory frame would indicate the spatial electromagnetic wave energy flux vector projected on the spatial trajectory of the corpuscle electron in the laboratory frame. Because $\mathbf{v_t}$ and $\mathbf{k}$ are vectors in the laboratory frame, the velocity's orientation of the corpuscle electron in the laboratory frame would correspond to the electromagnetic energy flux wave vectors, which is the Poynting vector.

While passing through an external potential of an atom (space scale of 10^{-11} m), the direction and intensity of the interaction between the atom's electromagnetic field and the electromagnetic field of the internal corpuscle electron at a scale smaller than 10^{-11} m would rectify the frequency and contour, which is the rotation radius, $r_t = c/\omega_t$, of the circulating electromagnetic field of the vorticity photon, based on the interactive temporal duration $\delta t = t\text{-}t_0$. At this temporal scale ($\approx 10^{-18s}$), the spatial dimension is about 10^{-10} to 10^{-13} m, which matches the case for high magnification of a transmission electron microscope.

If we consider a corpuscle electron traveling in a region that has the electric field potential Λ, the energy of the corpuscle electron in this region is

$$\varepsilon_t = \gamma m_0 c^2 + \Lambda_t = h\omega_{t_1} + \Lambda_t = h\omega_{t_2}$$

$$h(\omega_{t_2} - \omega_{t_1}) = \Lambda_t = h\Delta\omega_t = hc\Delta k_t, \quad \Delta k_t = \Lambda_t/hc \text{ or } \Delta\omega_t = \Lambda_t/h$$

This indicates that at every locus point on the trajectory in this spatial area, the wave vector of the electromagnetic energy flux in the internal frame is modified by the electromagnetic field at the locus point on the spatial trajectory at the elapsed time, and/or a modulation of the elapsed frequencies of the circulating electromagnetic waves of the corpuscle electron. The electric potential field manipulates the electromagnetic wave vector direction, which is related to the variation of the frequencies, inducing the velocity deviation from the incident direction and frequencies modulated. If the interaction duration, δt, is shorter than the temporal period of the circulating photon ($<10^{-20}$s) and the spatial distribution ($<10^{-13}$ m) of the field symmetrically, the corpuscle electron will not change its moving direction. If the rectified frequencies have an asymmetric distribution, it would result in the deviation of the spatial trajectory of the corpuscle electron in the laboratory frame, $v_t = c^2/v_{ph}$ as shown in Figure 3.4.

For an electron microscope, a flying corpuscle electron with velocity v_t should transmit through the sample in a short time. If the sample thickness is from >1 nm to >100 nm and velocity of an electron is from 0.54c to 0.94c, the passing temporal duration is from 10^{-18} s to 10^{-16} s or from attoseconds to femtoseconds. As mentioned, the Zitterbewegung vorticity frequency is around ZHz ($\approx 10^{18}$–10^{-22} Hz), therefore for sample having thickness with a few nanometers and 100–200 kV acceleration electrons it is possible to interaction between the electromagnetic fields of the corpuscle internal field and the medium field during the corpuscle electron passing through the sample. For thicker sample (more than a few nanometers), the corpuscle electron would just be an energetic particle.

Due to the corpuscle electron being a confined assembly of kinetic electromagnetic energy flux of the circulating vorticity field of the photon in a vacuum, when the corpuscle electron transmits through a sample, its electromagnetic field has to propagate in the medium of the sample with new electric and magnetic parameters, ε_s and μ_s, which are spatio-temporal functions. For a crystal, the spatio-temporal function may be a periodic function of its spatial lattice or crystal momentum k as $\varepsilon_s = \varepsilon_s\,(x,y,z,t) = \varepsilon_0 \exp i\phi\,(k_j,\Omega)$. The electromagnetic fields in a crystal may be expressed as $D = \varepsilon_s\, E$ and $H = \mu_s\, B$, which are also periodic functions. The electromagnetic field distribution in a primitive unit cell will be translated periodically with the symmetries. Its corresponding reciprocal primitive unit cell, which is the Brillouin zone, is also translated periodically. There always exists a crossing zone of the Brillouin boundary, which usually is the symmetry center of the Brillouin zone and has minimum potential as an avoid point of space field or tunnel. The corpuscle electron can fly through the tunnel without deviation, but if the corpuscle electron's flying direction is not exactly aligned with the tunnel, which can be visualized as a tunnel of a grating, the flying corpuscle electron would be deviated from the incoming direction due to the asymmetric electromagnetic field distribution on its path of the trajectory in the Brillouin zone, as shown in Figure 3.8.

As previously discussed, the corpuscle electron is an energetic electromagnetic wave pulse with super high frequencies and velocity, $\omega_{\parallel}, v_{\parallel}$, in the parallel optical axis direction and with lower frequencies, $\omega_{\perp}$, and velocity, $v_{\perp}$, in the transversal direction of the optical axis, due to the relative Doppler effect. The elapsed duration of the corpuscle electron passing through the sample is dominated by the longitudinal velocity $t = \Delta z/v_{\parallel}$ (Δz is the thickness of the sample), and simultaneously, the transversal component of the circulating electromagnetic energy flux would interact with the sample's transversal field distribution, resulting in transversal momentum exchange and wave vector modulation exhibited in the variation of its propagation direction. The escaped flying corpuscle electron would carry the information of the interaction in its amplitude and phases.

3.5.2 Flying corpuscle electrons traverse through the sample in the spatial and temporal domains

Based on Newton's second law, the velocity is such that the applied force, f, balances the rate of change of momentum $f = d(mv)/dt$ and the energy or work expresses the fact that any increased energy originates from the work done, $d\varepsilon = f\ dx$, which can be expressed as the rate of work $d\varepsilon/dt = fv$. If we use $\varepsilon = mc^2$ this, from the above equations we can have the following identity:

$$m(d\varepsilon/dt) - (mv)\,d(mv)/dt = c^2 d\left\{\left[m^2\left(1-(v/c)^2\right)\right]\right\}/dt = 0$$

From this identity, we may derive the relationship between mass and velocity, which $m(v) = m_0[1 - (v/c)^2]^{-1/2}$ is the well-known formula in Einstein's theory of relativity. However, it is now deduced from Newton's second law.

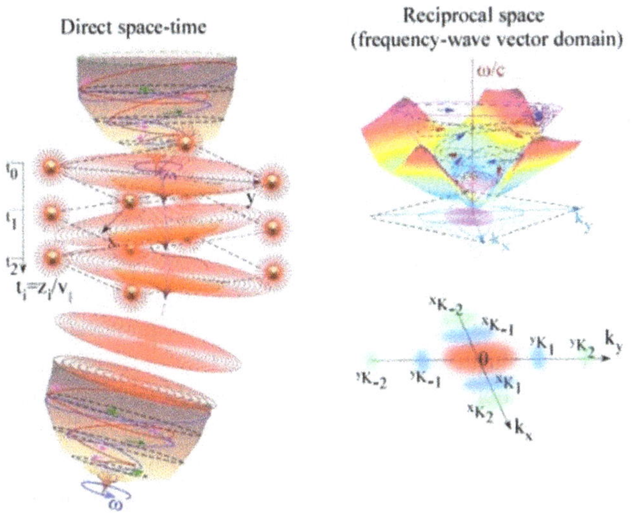

Figure 3.8: The corpuscle electron flying through the tunnel or avoiding the point of symmetric center of the Brillouin zone of a square lattice.

The mass of a system is not a constant, but it is a variable characteristic of the mass of a system. For the electron, as a corpuscle of confined electromagnetic energy, its total energy is

$$\begin{aligned}\varepsilon_{tot} &= m_0 v^2/(1-v^2/c^2)^{1/2} + m_0 c^2 (1-v^2/c^2)^{1/2} \\ &= mv^2 + m_0 c^2 (1-v^2/c^2)^{1/2}\end{aligned}$$

This relation is called the Planck-Laue formula, which unveils the relationship between the kinetic and dynamic energy of the corpuscle electron that plays a key role in modern physics.

If the kinetic energy of a moving corpuscle electron is ε_k, then

$$\varepsilon_k = mv^2 = v^2/c^2 (mc^2) = v^2/c^2 \varepsilon_{tot}$$

That means the kinetic energy of an electron is partial energy of the electron due to $v/c < 1$, as we already discussed in the previous part.

If the velocity of the electron, v, in the microscope frame is less than the speed of light, c (i.e., $v < c$), the kinetic energy would be as

$$\varepsilon_k = \varepsilon_{tot} - m_0 c^2 = mv^2 - m_0 c^2 \left[1 - (1 - v^2/c^2)^{1/2}\right]$$

If the velocity of the electron is $v \ll c$, then we may expand the roots to get

$$\varepsilon_{tot} = \varepsilon_k + m_0c^2 = 1/2m_0v^2 + m_0c^2$$

That tells us that the moving energy of a corpuscle electron only constitutes part of the total energy, and the internal energy is the fundamental energy of the corpuscle, which means the electromagnetic field energy is the inherent or intrinsic energy. It is important to always remember that the motion of a corpuscle electron is in the inertial frame, such as a laboratory or microscope frame, but simultaneously, the internal circulating electromagnetic field of the electron is in dynamic motion within its internal inertial frame of the corpuscle as a harmonic oscillator. That tells us if we measure the characteristics of the circulating electromagnetic energy in the microscope frame, you have to consider the relativistic transformation of the two inertial frames.

In an electron microscope, the motion of the electron is in the laboratory frame, but if you consider the high magnification, for example, 10^4, you may think about the relativistic transformation between the observer and the internal frame of the corpuscle.

It is significant to understand the physical nature of the de Broglie wave and the internal dynamic process of the corpuscle, which is a confined electromagnetic energy. If we accept the kinetic energy of a corpuscle as

$$\varepsilon_k = \varepsilon_{tot} - m_0c^2/\left(1 - v^2/c^2\right)^{1/2}$$

In an electron microscope, the flying corpuscle electron is always around the optical axis with a small cone angle α, as shown in Figure 6.5. So the velocity v may be expressed $v = v_{||} + v_{\perp}$ and $(v/c)^2 = \left(v_{||}/c\right)^2 + (v_{\perp}/c)^2$ and using Einstein's relation $\varepsilon = h\omega$ and $\varepsilon_k = v^2/c^2\varepsilon_{tot}$. As discussed in Chapter 1, the circulating electromagnetic energy flux has dynamic motion on periodic paths of the trajectory, which is an S(2) × S(2) topological surface. The kinetic velocity has a forward half-period and a rearward half-period on its trajectory with a vorticity field in the corpuscle electron frame, which has a center of rotation as the stationary center of mass at the rest state of the corpuscle electron. If we use the velocity, v_t, of the corpuscle electron as the velocity of the mass center of the corpuscle electron in the corpuscle electron frame, the rotation frequencies of the vorticity field of the circulating photon are

$$\omega_1 = \gamma\omega_0(1 + v/c)\ \ \omega_2 = \gamma\omega_0(1 - v/c)\ \ k_1 = \gamma k_0(1 + v/c)\ \ k_2 = \gamma k_0(1 - v/c)$$

The electromagnetic energy carrier, which is the circulating vorticity field of the photon, has the velocity of the electromagnetic energy carrier, $v_{\mathcal{E}} = \mathbf{S}(k)/\mathcal{E}$, where $\mathbf{S}(k)$ is the Poynting vector and $\mathcal{E}$ is the electromagnetic energy density:

$$v_{\mathcal{E}} = (\omega_1 - \omega_2)/(k_1 + k_2)\ \ \gamma = \left(1 - v^2/c^2\right)^{-1/2} = (\omega_1 + \omega_2)/2\omega_0$$

The de Broglie matter wave is the kinetic energy of the electromagnetic energy carrier in the corpuscle electron frame. The de Broglie matter wave then has the frequency and wave vector as

$$\omega_{db} = (\omega_1 + \omega_2)/2 \quad k_{db} = (k_1 - k_2)/2$$

Based on Einstein's relation $\mathcal{E} = h\omega = mc^2$, for the de Broglie matter wave, we have

$$\varepsilon = h\omega_{db} = h[(\omega_1 + \omega_2)/2] \quad p = hk_{db} = h[(k_1 - k_2)/2]$$

The corpuscle electron may be described by the electromagnetic energetic particle or pulse with forwardly and rearwardly moving waves on a closed spatio-temporal loop. Then, based on the relative equation of motion, $\mathcal{E}^2 - p^2 = m^2$, we may have

$$\omega = (\omega_1\omega_2)^{1/2}$$

As shown in Figure 3.5, which

$$v/c = (\omega_k/\omega_{tot})^{1/2} = (\omega_k\omega_{tot})^{1/2}/\omega_{tot}$$

If $\omega_v = (\omega_k\omega_{tot})^{1/2}$, then we may have $v_t = c(\omega_t/\omega_{tot}) = c(k_t/k_{tot}) = c\tan\theta$, where θ is the electromagnetic field momentum, Poynting vector, inclined angle with velocity. As shown in Figure 3.5, $\omega_{tot} = \omega_{||}$, which is Doppler blue shift and the $\omega_t = \omega_{\perp}$, which is Doppler red shift, and $k_{tot} is\ k_{||}$ That relation clearly shows the flying corpuscle electron around the optical axis of the microscope has the super high frequencies pulsating characteristics of electromagnetic energy flux. The temporal and spatial interval between the flying corpuscle electrons would correspond to the additional temporal and spatial delayed frequencies for the flying corpuscle electron.

The de Broglie's wave phase describes the dynamic periodic motion of the electromagnetic energy flux in the internal frame of the corpuscle electron. If we accept this view, the phase between the field and the dynamic internal motion of the electromagnetic field of the corpuscle electron should be identified. If there is no external interaction, for example, electric and magnetic fields, it means the phase should be equal for the observer frame and the internal frame of the moving corpuscle electron. Then the phase

$$\begin{aligned}\phi &= 1/\hbar(\varepsilon_v t - px)\\ &= 1/\hbar\left[\varepsilon_0/(1-(v^2/c^2))^{-1/2}\right]\ [t-(v/c^2)vt]\\ &= (1/\hbar)\left[\varepsilon_0/(1-(v^2/c^2))^{-1/2}\right](1-v^2/c^2)t\\ &= (\varepsilon_0/\hbar)\ \left[(1-v^2/c^2)^{1/2}\right]t = (\varepsilon_0/\hbar)t' = \omega_0 t'\end{aligned}$$

where $k = 1/\lambda = p/h = m_0 v/h(1-v^2/c^2)^{1/2}, t = t'(1-v^2/c^2)^{-1/2}$, this formula connects a particle's (corpuscle) momentum in the microscope frame with its phase of the electromagnetic energy wave in the corresponding internal frame of the corpuscle electron. The phase is the rotation angle difference corresponding to the temporal dilation between the corpuscle electron's internal frame and the microscope frame. For a flying corpuscle electron flow or beam, the phase of an individual corpuscle electron would have the temporal interval between the consecutive flying corpuscle electron pulses, resulting in a frequency difference $\Delta\omega$. If the temporal interval is constant and stable, then the flying corpuscle electron is in complete coherence.

At higher magnification of a transmission electron microscope, the spatial scales observed on the screen in the laboratory frame are $>10^{-6}$ mm or 10 nm for 1 mm on the screen. The de Broglie wavelength in the transversal dimension is about 2 nm with divergence $\alpha \approx 10^{-3}$ rad and 20 nm with $\alpha \approx 10^{-4}$ rad at 300 kV acceleration voltage. This means the physical process of the corpuscle electron may be considered as occurring in its internal frame instead of the laboratory frame, similar to a space shuttle flying near Mars. Because $s = vt$, the temporal scale also changes to $\tau = 10^{-7\ s}$ or ns (nanosecond). Therefore, the spatio-temporal domain changes from low magnification to higher magnification in the microscope. This dimensional change leads to understanding the corpuscle electron from an energetic particle to a corpuscle electron, which is a confined, rotated, circulating electromagnetic wave flux enveloped in the corpuscle's own inert frame. This means the atom's spatial configuration of a sample overlaps with the corpuscle's internal spatio-temporal domain at similar scales, and the confined rotating electromagnetic energy vortices field may be traveling through the atom's configuration space. This is the meaning of the de Broglie matter wave in electron microscopy at higher magnification.

The moving corpuscle electron with velocity v may be expressed as a function such as

$$\Psi(r,\ t) = \Psi_v \exp((i/\hbar) m_0 c^2 \tau)$$

in which

$$\Psi_v = E_0(\delta(r - r_0)\delta(t - \tau_0)) \exp[i/\hbar(pr - \varepsilon_v t)]$$

Then

$$\Psi(r,t) = \{E_0(\delta(r - r_0)\delta(t - \tau_0)) \exp((i/\hbar) m_0 c^2 \tau)\} \exp[i/\hbar(pr - \varepsilon_v t)]$$

It may be shown that the photon with electric field E_0 has harmonic motion with frequency $\omega_0 = m_0 c^2/h$ in the electron's flying direction. The frequency and phase of this harmonic motion are modulated by the corpuscle's kinetic motion such that exp[i/ħ (pr – $\mathcal{E}_v$t)]. The modulated frequency depends on the energy difference between the instantaneous energy $\mathcal{E}_t$ and the rest energy $\mathcal{E}_0$ of the corpuscle electron in the laboratory frame. The phase modulation is related to the spatial environment character of the rotating and circulating photon. The modulation wave is the envelope function that confines the harmonic motion of the electromagnetic wave photon. It is usually called the de Broglie wave. The function $f(z,\tau) = E_0(\delta(z - v_z \tau))\exp(-(i/\hbar) m_0 c^2 \tau)$ is the energy carrier of the harmonic motion of the circulating electromagnetic flux, and the energy carrier rotating around the z-axis may present as $\psi_0 = f(z - v_r t)\exp\{pr - \omega_0 t\}$, in which v_r is the velocity of electromagnetic energy flux rotation in the x-y plane and ω_0 is the rotating angular frequency of the energy carrier around the z-axis. When the corpuscle starts to move along the z-axis with velocity v_t, then relativity theory has to be considered. The wave function may be given as

$$\psi\{x, y, z, t\} = f\{x, y, \gamma(z - v_t t)\}\ \exp\{i\gamma\omega(t - v_t z/c^2)\}$$

where the energy carrier is f {x,y, $\gamma(z-v_t t)$} and is moving at velocity v_t, indicated by the Lorentz factor γ, which means the contraction of the length due to special relativity. The phase of the electromagnetic harmonic oscillation wave is expressed as exp {$i\gamma\omega(t-v_t z/c^2)$}, which is a transversal wave moving with the carrier wave at the velocity c^2/v_t; the frequency ω is identified as $\omega_0 = m_0c^2/h$, but mass is $m = \gamma m_0$. That implies $\omega = \gamma\ \omega_0$. The de Broglie wave vector or wave number is $k_{db} = \gamma\ \omega_0\left(v_t/c^2\right)$ and the de Broglie wave of a moving corpuscle electron with velocity v_t may be expressed as

$$\psi\{x, y, z, t\} = f\{x, y, \gamma(z - v_t t)\}\exp\{-i(k_{db}z - \omega_{db}t)\}$$

It is clear that the corpuscle electron is (i) the electromagnetic energy carrier, (ii) the electromagnetic harmonic oscillator, and (iii) the electromagnetic energy varies with the phase of the spatio-temporal domain. When $\varphi = k_{db}$ z-ωt = 0, then ψ {x, y, z, t} = f {x, y, $\gamma(z-v_t t)$} and $d\psi/dt = df/dt = v_t$,which expresses the velocity of the corpuscle electron. For the moving corpuscle electron, the phase is equal to zero, which means $v_{ph} = v_z$ due to $v_{ph} = \omega/k_{db}$ and $z/t = v_z$. That indicates that the locations with phase equal to zero have the identical velocity between the internal harmonic oscillation's phase and the moving corpuscle. In other words, at these locations, the corpuscle electron may act as a harmonic oscillator of the electromagnetic field, which has an invariant phase $\varphi = k_{db}$ z–ωt. The phase value would be the same in the laboratory or in the internal frame. On the wave front of the propagating wave, there are two modes: (1) spatial mode, which implies the electromagnetic field oscillates at the spatial locus and the oscillation intensity decays exponentially and (2) temporal mode, which means the oscillation is perpetually harmonic at the location $z_t = v_t t$. The phase $\varphi = k_{db}$ z–ωt contains these two modes in which $\omega = \gamma\ \omega_0$ expresses the oscillation frequency and $k_{db}z = \omega_0(z/c) = \omega_0 t$ is the temporal mode. The wave function contains the information.

3.5.2.1 From single electron to multielectron beam

The isolated corpuscle electron should be mathematically described as a function, usually called a wave function $\psi = F_0(x, p)\exp(i/h(\nu t - kx))$ The wave function exhibits the kinetic and dynamic process of the electromagnetic energy flux in its internal inertial frame, which is a periodic phenomenon expressed as exp(i/h(νt – kx)), in which $k = 1/\lambda = 2\pi\, m_0 v\, /h(1 - v^2/c^2)^{1/2}$and $\nu = \nu_0(1 - (v/c)^2)^{-1/2}$. These equations indicate the relationship between the velocity of the corpuscle electron in the laboratory frame and the phase of the electromagnetic wave field in the internal frame of the corpuscle. It is well known that the intensity of light is given by $I(r, t) \sim |E(r,t)|^2$ which implies the number of photons passing through a cross-section per unit time. The $|F_0(x, p)|^2$ may be defined as the energy intensity distribution of the corpuscle electron. Therefore, if we use $\psi_0(x.t) = \exp(i/h(\nu t - kx))$ only the periodic motion, the $|\psi_0(x,t)|^2$ is the energy density of a corpuscle electron to be detected at location x and time t. This $\int_v \psi\psi^* dv = |F_0(x,p)|^2$ expresses that there exists a corpuscle electron having energy density $|F_0(x,p)|^2$ in the volume v. As discussed in the previous section, the energy density of a corpuscle electron is the electromagnetic pulsating energy with a super high frequency. Therefore, the wave function may switch to the function of the electromagnetic field. Multiple corpuscle electrons fly successively on spatial trajectories having different spatial frequencies $K = 1/\delta s$ with a time delay $\delta t = \delta s/v_t$. The corpuscle electron beam may be described as an ensemble of electromagnetic field pulses, where a single temporal pulse realization taken from the successive flying corpuscle electron can be expressed with a complex analytic pulse field $E(\mathbf{r},t)$, where $\mathbf{r} = (x,y,z)$ is the spatial position vector in three-dimensional coordinate space and the field propagates toward the positive half-space $z > 0$. This forms a Fourier transform pair with the spectral field realization:

$$E(\mathbf{r},\omega) = (1/2\pi)\int_0^{T0} E(\mathbf{r},t)\exp\{i\omega t\}dt \quad E(\mathbf{r},t) = \int_0^{\infty} E(\mathbf{r},\omega)\exp\{-i\omega t\}\,d\omega$$

The field $E(\mathbf{r},t)$ square integrable as $\int |E(\mathbf{r},t)|^2 dt < \infty$ is satisfied. This requirement means the electric field realization taken from the flying corpuscle electron train is not infinite in the temporal domain, but rather it consists of pulses. The complete train of corpuscle electron pulses, $\tilde{E}(\mathbf{r},t)$, is reconstructed when these single realizations are arranged in a line along the propagation direction as $E(\mathbf{r},\tau) = \sum^{N}_{m=0} E_m(\mathbf{r}, t + m\Delta\tau_m)$ where the pulse separation $\Delta\tau_m$ and each realization $E_m(\mathbf{r},t)$ are potentially different in a shot-to-shot manner, i.e., the train may fluctuate in both the shape and timing of the pulses. When the corpuscle electron pulse train is completely stable, the pulses are all identical and equally spaced. Importantly, the corpuscle electron pulse train $\tilde{E}(\mathbf{r},t)$ forms a Fourier transform pair with

$$\tilde{E}(\mathbf{r},\omega) = \sum_{m=0}^{M} E_m(\mathbf{r},\omega)\exp\{i\omega m\Delta\tau_m\} \quad \tilde{E}(\mathbf{r},\tau) = \sum_{m=0}^{N} E_m(\mathbf{r}, t + m\Delta\tau_m)$$

When all of the corpuscle electron pulses' realizations $E_m(\mathbf{r},\omega)$ are identical and $\Delta\tau_m$ is constant, the corpuscle electron train becomes completely coherent, and it can be written as

$$\tilde{E}(r,\omega = E(r,\omega)D_M(\omega,\Delta\upsilon)$$

where $D_M(\omega, \Delta\upsilon)$ is a Dirichlet kernel:

$$D_M(\omega,\Delta\upsilon) = \{\sin[(m+1)\omega\pi/\Delta\upsilon]/\sin(\omega\pi/\Delta\upsilon)]\}\exp(M\omega/\Delta\upsilon)$$

with $\Delta\upsilon = 2\pi/\Delta\tau$ being a constant separation between spectral components. This ideal identity is called a frequency comb. When the number of pulses tends toward infinity, the Dirichlet kernel takes the form of a perfect delta comb $D_\infty(w,\Delta\upsilon) = \Sigma\,\delta(\omega - m\Delta\upsilon)$.

For TEM, the frequency, ω, of a corpuscle electron pulse depends on the magnitude and orientation of the velocity of the flying corpuscle corresponding to the optic axis of the microscope and the separation between spectral components dominated by the emission method (for example, thermal emission, field emission, and laser-radiated carbon nanotube). The electron beam as a corpuscle electron pulse train may be expressed as

$$\tilde{E}(r,\omega) = E(r,\omega)D_M(\omega,\Delta\upsilon) = \sum_{m=0}^{N} E(r,\omega)\delta(\omega - m\Delta\upsilon).$$

But if we consider the internal dynamic electromagnetic field of the corpuscle electron, the electron beams may be expressed as

$$\begin{aligned}\tilde{E}(r,t) &= |E(r,t)|\Sigma\delta(\varphi)\delta(z - v_t t)\delta(\omega - 2\pi m/\Delta\tau)\\ &= |E(r,t)|\exp\{i\psi(t)\}\exp\{i\psi_0\}\Sigma_m\delta(z_m - v_t t_m)\delta(\omega - m/\Delta\tau)\end{aligned}$$

where $|E(r, t)|$ is the space-time dependent envelope, ω is the carrier frequency, $\psi(t)$ is the time-dependent phase, and ψ_0 is a constant known as the carrier-envelope offset phase. The square of the envelope, $I(\mathbf{r}, t) = |E(\mathbf{r},t)|^2$ is the spatio-temporal-dependent instantaneous power of the corpuscle electron pulses that can be measured if a detector of sufficient bandwidth is available. For an electron microscope, the spatio-temporal-dependent instantaneous power of the corpuscle electron pulses can be exhibited at two planes: (a) at the plane beyond the sample where the single corpuscle electron just escaped from the sample, also called the near-field, and simultaneously this plane is located near the front focus plane of the objective lens, which is adjustable by defocus of the lens and (b) the plane located at the back-focus of the objective lens, also called the far-field domain. The near-field region contains the electromagnetic field intensity redistribution, which relates to the revived spatial periodicity of the atoms grating of the sample (or Talbot effect), encoding the history of suffering (or chirping) of the internal circulating photon in the medium of the sample, for which the confined electromagnetic field restarts to propagate in a vacuum again. That repartitioning density distribution revives the "spatial periodic objective function," E(x, y, z), for the imaging of the objective lens. The "back-focus plane" of the lens is the far-field domain or Fraunhofer field, which is the electromagnetic energy wave front with suffering history propagating for a long time and smearing out the spatial periodic objective information, instead only having the angular distribution of the near-field intensity, but without any spatial and temporal information of the interaction between the atoms of the sample and the interacting corpuscle electron. That tells us if we separate the function of the corpuscle electron pulses train, $\tilde{E}(r, t)$, into spatial and temporal parts, the $\tilde{E}(r, t)$ function is the convolution of the spatial and temporal distribution of the corpuscle electron pulses trains. We can separately measure the spatial characteristics (position and angle) and temporal features (angular frequency) at the near-field region and only the angle distribution of the corpuscle electron pulsed train at the far-field domain for TEM. Using the laser beam with appropriate frequencies interacting with the escaped repartitioned corpuscle electron pulses train at the near-field region may collect the spatio-temporal information of the interacted atoms of the sample that should be developed.

3.5.2.2 Flying corpuscle electron self-interference and beam coherence

The flying corpuscle electron beam as $\tilde{E}(r,t) = |E(r,t)|\Sigma\delta(\varphi(\mathbf{r},t))\delta(z - v_t t)\delta(\omega - 2\pi m/\Delta\tau)$ is flowing with a temporal separation $\Delta\tau$, which is the time in the microscope frame, and an isolated corpuscle electron with e/m and 0.5 h spin having velocity v_t is also in the microscope frame. The delta comb, $D_\infty(\omega, \Delta\upsilon) = \Sigma\delta(\omega - m\Delta\upsilon)$ expresses the continuous pulses flowing with the same frequency, ω, which means the same energy hω for the corpuscle electron and frequency interval $\Delta\upsilon = 1/\Delta\tau$. The stability of a flying corpuscle electron train can be characterized by the mutual coherence function (MCF), Γ, in the temporal domain and the cross-spectral density (CSD) function, $\hat{W}$, in the spectral domain:

$$\Gamma(r_1, r_2, t_1, t_2) = \langle E^*(r_1, t_1)E(r_2, t_2)\rangle = (1/N)\sum_{N=1}^{N}(E_n^*(r_1, t_1)E_n(r_2, t_2)$$

$$\hat{W}(r_1, r_2, \omega_1, \omega_2) = < E^*(r_1, \omega_1)E(r_2, \omega_2) >$$

The angle brackets denote averaging over an ensemble of single corpuscle realizations picked from the flying corpuscle train. The single corpuscle realizations are picked from the train with a moving sampling window, with a period that is equal to the average value of the corpuscle separation. It is possible to also account for random variations of Δτ (as timing jitter) as well as any changes in the corpuscle shape or phase. In the temporal domain, ensemble averaging over long sections of the pulse train would lead to a correlation function that repeats. The mean intensity I(r,t) and spectral density $\hat{S}(r,\omega)$ can be used to normalize the MCF and the CSD, which yields the complex degrees of temporal and spectral coherence:

$$\gamma(r_1, r_2, t_1, t_2) = \Gamma(r_1, r_2, t_1, t_2)\ /[I_1(r_1, t_1)I_2(r_2, t_2)]^{1/2}$$

$$\eta(r_1, r_2, \omega_1, \omega_2) = \hat{W}(r_1, r_2, \omega_1, \omega_2)/[\hat{S}_1(r_1, w_1)\hat{S}_2(r_2, w_2)]^{1/2}$$

For TEM, the MCF is related to the emission process of the electron, the electron gun stability, and the angle divergence of the velocity of the electron at the electron crossing point of the electron gun. The interference pattern of the divergent corpuscle electron beam with a field emission gun is formed by the interference of the confined electromagnetic dynamic wave fields of multiple single corpuscle electrons with the same energy because the MCF is equal to 1.

3.5.2.3 Electron beam traverses a sample

As previously discussed, the flying corpuscle electron beam flows $\tilde{E}(r,t) = |E(r,t)| \Sigma_m \delta(\varphi(r,t))\delta(z_m - v_t t_m)\delta(\omega - 2\pi m/\Delta\tau)$ with a temporal separation $\Delta\tau$, which is the temporal separation of an isolated corpuscle electron having velocity v_t in the microscope frame. When the flying corpuscle electron trains travel through a sample, every single corpuscle electron pulse has to endure the electromagnetic field of the sample medium, resulting in energy exchange and dynamic phase modification in the spatial and temporal domains. After this process, the corpuscle electron pulse trains have different individual single corpuscle electron realizations with modified spatial and temporal frequencies. These single corpuscle electron realizations may couple or pair with each other, causing near-field mutual coherence interference and far-field Fraunhofer diffraction, resulting in the image and diffraction in TEM.

In TEM, the plane wave is often used for an electron beam, but it is unclear how a charge could spread to the whole space. Originally, using an electron beam as the illuminating source was based on the charged particles' current without a clear idea of the wave-particle duality of the electron. Now, the electron is a structured, complex, physical entity, in which the electromagnetic field, circulating and rotating harmonically in two perpendicular spatial planes, constructs the confined electromagnetic energy flux corpuscle electron that possesses particle behavior (energy and momentum) and wave peculiarity (frequency, wavelength, and phase). The internal harmonic motions would be modulated by the corpuscle's velocity in the observer frame due to relativity effects. The energy carrier is the internal rotating and circulating electromagnetic harmonic photon, but the observed high-energy electromagnetic waves with the Doppler effect are the de Broglie waves, which are manipulated by the velocity of the corpuscle electron. For the electron microscope frame, the corpuscle electron motion as super high-frequency pulse trains should be consecutively flying. Due to the de Broglie relationship $\mathbf{p} = h\mathbf{k}$ and velocity divergence $\alpha = p_r/p$, $p^2 = p_r^2 + p_z^2 = p_x^2 + p_y^2 + p_z^2$, when the corpuscle electrons fly consecutively along the optical axis (or z-axis) with a divergent angle a $<10^{-3}$ rad, the ratio of the longitudinal and transversal de Broglie wavelengths is larger than 10^3. Therefore, the distribution contours of the electromagnetic field energy of the corpuscle electron in the microscope frame look like a flat disk, which has a larger scale in the transversal direction and a small dimension in the longitudinal direction, and that could be measured by the impulsive momentum retrieval method shown in Figure 3.10.

As previously mentioned, the v_t/c of a flying corpuscle electron in the z direction may be expressed as follows: $v_t/c = (\omega/\omega_0)(r/r_0) = rp/h$ and $p = v_t h/cr = (h/c)(v_t/r)$. The p and v_t are vectors, and p_x would relate to v_x. Then, we may have $p_x = (h/c)v_x/r_x = (h/c)(1/\tau)$. It tells us that the transversal momentum of a flying electron would have an inverse proportion with the kinetic temporal periodicity τ_0 of the vorticity photon in the corpuscle electron frame, which corresponds to the time $\tau = \tau_0(1-(v_x/c)^2)^{-1/2}$ in the laboratory frame. However, due to $\omega^2 = \omega_0^2 + (ck_{db})^2$, the $(\omega + \omega_0)(\omega - \omega_0) = c^2k^2$ Doppler effect, the blue shift occurs in the corpuscle's moving direction, and the red shift occurs in the transversal direction. Therefore, the p_x has a smaller temporal interval.

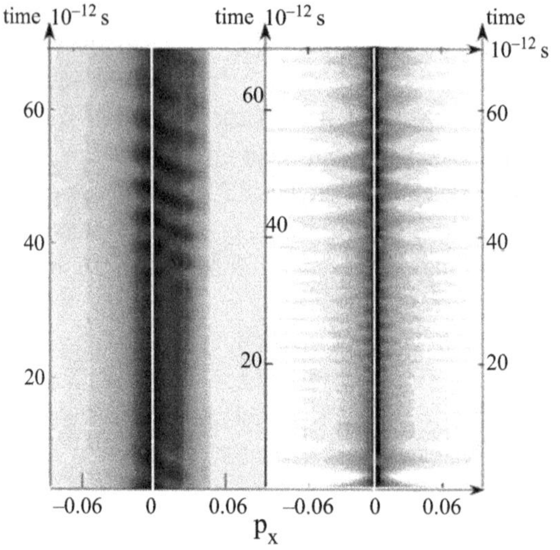

Figure 3.10: The experimental momentum p_x distribution (left) and its simulation (right), which shows the corpuscle electron flying consecutively with a time interval $\Delta\tau$ as a pulse train.

From the experimental result, we believe that the interactions between the electron beam and a sample involve an elementary process experienced by a single individual corpuscle electron in the medium of the sample. This means that the electromagnetic fields of a single corpuscle electron are influenced by the electromagnetic fields at the loci of its paths in the medium of the sample due to the Lorentz force, causing the modification of dynamic parameters (k, vt, ω) of the circulating electromagnetic energy flux of the rotating vorticity field of a photon, resulting in the rectification of the electromagnetic energy flux velocity, which is the energy flux velocity $v_\varepsilon = \langle S_k \rangle_{t,s} / \langle \varepsilon_k \rangle_{t,s}$ (S_k is the Poynting vector). The interaction process occurs within the x-y-dimensions of the sample and the time duration $\Delta t = \Delta z / v_t$ (traveling time in the sample). As discussed previously, space and time must be fused with each other, and electromagnetic wave propagation implies that the wavefront with the same electromagnetic harmonic oscillator moves with velocity v_t, which is the ratio of space and time as $\delta s/\delta t$. Therefore, for an electron flying through the sample, the observer at the microscope frame should observe the spatial phenomenon, which is the intensity spatial distribution of accumulated corpuscle electrons flying from the sample to the observing screen during the recording time. As discussed previously, the MCF dominates the single corpuscle electron pulse having a delayed time, $\Delta\tau$, with the same spatio-temporal frequencies that can couple with each other to form a stable spatial interference pattern when the corpuscle electrons with delayed time, $\Delta\tau$, have crossed over each other at the same spatial point due to the phase difference between the crossing corpuscles. The observed interference pattern of the corpuscle electron beam is the same interference pattern of two single crossing corpuscle electrons, but it is weighted by the multiplying accumulated numbers of the corpuscle electrons. In the laboratory frame, these weighted interference patterns exhibit the periodic distribution of electron density in physical space.

3.5.2.4 Single corpuscle electron moves with velocity v_t

For a coherent corpuscle electron beam or ensemble, it is possible to use the characteristics of a single corpuscle to express the beam's behavior because every realization of the corpuscle electron pulses is the same. Therefore, a single corpuscle electron can be described as

$$\tilde{E}(r,t) = |E(r,t)| \sum \delta(\varphi)\delta(z - v_t t)$$
$$= |E(r,t)| \exp\{i\psi(t)\}\exp\{i\psi_0\}\exp\{i\omega t\}\delta(z - v_t t)$$

The derivative of the time-dependent phase, $\psi(t)$, accounts for the occurrence of different frequencies at different times, i.e., $\partial\psi(t)/\partial t$ is the instantaneous frequency of the corpuscle electron pulse that describes the oscillations of the electric field around that time. The frequency representation of the corpuscle electron is defined by the Fourier transform:

$$\tilde{E}(r,\omega) = |E(r,\omega)| \exp[i\phi(\omega)] = \int_{-T}^{T} dt E(r,t)\exp(i\omega t)$$

where $E(r,t) = \int_0^\infty (d\omega/2\pi)\,\tilde{E}(r,\omega)\exp(-i\omega t)$ and spectrum intensity $I(r.\omega) = |\tilde{E}(r,\omega)|^2$. The spectral phase describes the relative phase of the transient frequencies composing the electromagnetic energy pulse, and its derivative $\partial\phi/\partial\omega$ is the group delay, $T(\omega)$, at the corresponding frequency, i.e., the time of arrival of a subset of composed frequencies of the electromagnetic energy pulse, ω. An electromagnetic pulse with a constant group delay, i.e., a linear spectral phase, is said to be Fourier transform limited because it is the shortest pulse that can be obtained for a given electromagnetic spectrum. The electromagnetic pulse is also characterized by the function $\tilde{E}(r,\omega)$ on the domain $[\omega_0\ \Omega, \omega_0 + \Omega]$, where Ω is a frequency that is large compared with the bandwidth of the circulating electromagnetic vorticity field (i.e., large compared with the inverse of the coherence time of the corpuscle electron).

A knowledge of the two-time correlation function allows one to determine whether the electromagnetic energy pulses in the ensemble are coherent, that is, to determine whether they have the same electromagnetic pulse field. A useful number characterizing the similarity of the electromagnetic pulses in the ensemble is the degree of temporal coherence μ:

$$\mu = \left(\iint |<E(t_1)E^*(t_2)>|^2 dt_1 dt_2\right) \Big/ \left(\int <E(t)\,E^*(t)> dt\right)^2$$

When this number is unity, all corpuscle electron pulses are the same, and values smaller than one indicate various degrees of statistical variations in the corpuscle electron pulse ensemble. In the case of identical corpuscle electron pulses, the correlation function factorizes, and the ensemble may be characterized by a single corpuscle electron pulsed field.

A two-frequency correlation function is the double Fourier transform of its temporal counterpart $Ç(\omega, \omega') = <\tilde{E}(\omega)\tilde{E}^*(\omega')>$, but it is better to write it in terms of cen ter and difference-frequency variables:

$$Ç(\Delta\omega, \omega_c) = <\tilde{E}(\omega_c + \omega t/2)\tilde{E}^*(t_c - \Delta t/2)>$$

or

$$C(t_c, \Delta t) = <\tilde{E}(t_c + \Delta t/2)\tilde{E}^*(t_c - \Delta t/2)>$$

where $\omega_c = (\omega + \omega')/2$ $\Delta\omega = (\omega - \omega')$, $t_c = (t + t')/2$ and $\Delta t = (t - t')$. measuring the correlation function is to make repeated measurements of the electromagnetic energy of the individual corpuscle electron that makes up the realization of the ensemble of accumulated received corpuscle electrons on the CCD device, which is related to the contrast of the image of the micrograph. The output of all absorptive detectors is proportional to a bilinear function of the corpuscle electron energy and a linear function of the two-time correlation function. The two-dimensional chronocyclic space (t,ω) can exhibit the intuitive concept of time-dependent frequency, which may be related to the frequency variation resulting from the flying corpuscle electron through the sample.

One approach to defining a representation of the pulse in the chronocycle phase space is a Fourier transformation of the correlation function with respect to the time difference of the two arguments:

$$\ddot{w}(t, \omega) = \int dt' <E(t + t'/2)E^*(t - t'/2)> \exp(i\omega t') \text{ or}$$

$$\ddot{w}(t, \omega) = \int (d\omega'/2\pi) <\tilde{E}(\omega + \omega'/2)\,\tilde{E}^*(\omega - \omega'/2)> \exp(i\omega' t)$$

The function $\ddot{W}(t,\omega)$ is known as the "chronocycle" Wigner function. Particularly useful characteristics of the Wigner function are that it is real-valued and that the marginals are the temporal and spectral intensities.

$$I(t) = |E(t)|^2 = \int (d\omega/2\pi)\ddot{W}(t, \omega) \quad \tilde{I}(\omega) = |\tilde{E}(\omega)|^2 = \int dt\ddot{W}(t, \omega)$$

The Wigner function is sufficient to characterize both individual corpuscle electron pulses and partially coherent corpuscle electron pulse ensembles. It is not positive definite in general and cannot be considered a probability distribution of the electromagnetic field of the corpuscle electron pulse. Negative Wigner functions are quite common even for simple corpuscle electron shapes and characterize many of the complicated pulse shapes. It should be noted that the restrictions on the pulse duration and bandwidth required by Fourier's theorem are inherent in the Wigner function, and there is a minimum area of the chronocycle phase space that it may occupy. The Wigner function acquires a slope indicating the correlation between time and frequency, and its contours provide some intuition about the pulse chirp via a graph of the time-dependent frequency.

As previously mentioned, if the flying corpuscle electron pulse train is coherent, then the beam behavior is the same as a corpuscle electron pulse with a constant time delay. The corpuscle electron pulse field may be expressed as

$$E(r,t) = |E_0(r,t)|\exp[i(k_0 z - \omega_0 t)]\psi(r, z, t)$$

The envelope function takes the beam profile form.

$$\psi(x,y,z,t) = \iint dk_x d\Omega \Psi(k_x, k_y, \Omega)\exp(k_x x)\exp(k_y y)\exp[i(k_z - k_0)z]\exp(-i\Omega t)$$

where $\Omega = \omega - \omega_0$ is the temporal frequency with respect to the corpuscle frequency ω_0. The spatio-temporal spectrum $\Psi(k_x, k_y, \Omega)$ is the three-dimensional Fourier transform of $\psi(x, y, 0, t)$ with respect to x, y, and t. For corpuscle electron pulse trains as discussed before, the de Broglie wave emanated from the rotation of the circulating electromagnetic energy flux of the vorticity field of a photon can be separated into rotating in the transversal space (or wave vector space) and propagating in the longitudinal direction as the temporal axis (or frequency domain) for a transmission electron microscope

$$\Psi(k_x, k_y, \Omega) \approx \Psi(k_x, k_y)\Psi(\Omega)$$

In which case $\Psi(x, y, 0, t) \approx \Psi_{x,y}(x, y)\, \Psi_t(t)$, where $\Psi(k_x, k_y)$ is the Fourier transform of $\psi_{x,y}(x, y)$ and $\Psi(\Omega)$ is the Fourier transform of $\psi_t(t)$. Besides the separability of the fields in space and time at $z = 0$, the fields remain approximately separable along the z-axis if the space-time coupling remains minimal, which requires narrow spatial and temporal bandwidths.

For a flying corpuscle electron train with constant energy, ω_0 is the frequency of the circulating electromagnetic energy of the vorticity field of a photon or the intrinsic frequency of Zitterbewegung, and ω_{dB} is the frequency of the dynamic rotation of the circulating electromagnetic energy flux of the vorticity field of a photon. The $\Omega = \omega - \omega_0$ may be seen as the de Broglie wave frequency ω_{dB}, which describes the frequency emanated from the kinetic motion of the corpuscle electron. As discussed previously, if $k_z = k_{||}$ and $[(k_x)^2 + (k_y)^2]^{1/2} = k_\perp$ and $k_0 = \omega_0/c$, then the fields of the flying corpuscle electron train and its envelope function may be written as

$$E(x,y,z,t) = \exp\{i(k_0 x - \omega_0 t)\}\psi(x,y,z,t)$$

$$\psi(x,y,z,t) = \int dkxdk_y \tilde{E}(k_x k_y)\exp\{i[(k_x x + k_y y)] - [(k_\perp)^2/2k_0]z)\}$$

The spatial spectrum $\tilde{E}(k_x, k_y)$ is the Fourier transformation of $E_x(x, y, 0)$. If the transversal field of the corpuscle electron train is periodic with interval period L, the spatial spectrum of the corpuscle electron train is discretized as

$$k_x \to nk_L = n(2\pi/L_x)\ k_y \to n(2\pi/L_y) \text{ and } \tilde{E}(k_x, k_y) \to \tilde{E}(nk_{L(x,y)}) = \tilde{E}_n$$

then

$$\psi_x(x,y,z) = \sum_n \tilde{E}_n \exp\{i[2\pi n((x/L_x + y/L_y) - n(z/z_T)]\}$$

n is an integer and $k_{L(x,y)} = 2\pi/L(x,y)$. Therefore, the initial contour of the corpuscle electron train is revived periodically at spatial Talbot planes $\psi_x(x, y, mz_T) = \psi_x(x, y, 0)$ for integer m with rich and striking rotating dynamics unfolding between these planes. But if the temporal envelope function $\psi(z, t)$ is expressed as an angular spectrum as

$$\psi(z,t) = \int d\Omega\psi_t(\Omega)\exp(-i\Omega t - z/\tilde{v})\exp(i1/2k_2\Omega^2 z)$$

where $\Omega = \omega - \omega_0, \omega_0$ is the center frequency of the ensemble corpuscle electron pulse train and Ω is the temporal frequency with respect to ω_0, $\tilde{v}$ is group velocity, $1/\tilde{v} = dk_z/d\Omega|_{\Omega=0}$, and $k_2 = d^2k_z/d^2\Omega|_{\Omega=0}$ is the group velocity delay (GVD). The temporal Talbot effect occurs when the field profile of the ensemble corpuscle electron pulse train at $z = 0$ is periodic in time with period T_t. The temporal spectrum of this corpuscle electron pulse train is discretized at $\omega = 2\pi m/T_t$ and the phase exp ($i1/2k_2\Omega^2 z$) responsible for the dispersive spreading takes the form $\exp(i2\pi m^2 z/z_{Tt})$. The pulse train first disperses, and the pulses overlap at temporal duration before it reconstitutes itself axially at planes separated by the temporal Talbot distance $z_{T_t} = T^2/\pi|k_2|$ with $\psi(\ell z_{T_t}, t) = \psi(0, t - \ell(z_{T_t}/\tilde{v}\cot\theta))$. The corpuscle electron pulse ensemble beam could be seen as ensembles of charge particle flow (electric current) but simultaneously also as the spatially and temporally revived confined dynamic rotating electromagnetic fields entity with periodic varying spatial profile and temporal haunting density of electromagnetic fields, which is the electron current. That means periodicity of spatial and temporal domains for a corpuscle electron has to cause the spatial and temporal Talbot effect, which results in the dynamic states of the electromagnetic fields of circulating photons at a corpuscle electron frame to be revived in space (as contour) and time (as propagating velocity) as a self-image. That may be the discretized quantum charge. At $r \geq (v_r/c)^2\ \lambda_{com}$ ($\lambda_{com} = 10^{-13}$ m) and $t > 10^{-15}$ s the corpuscle electron appears as a charge particle with e/m and 1/2 spin. For TEM to steer the component particle's velocity at the laboratory frame to induce the relative rotating phase difference of the dynamic electromagnetic fields in its internal frame. Due to the phase, $\phi = (\varepsilon_0\hbar/)\left[(1-v^2/c^2)^{1/2}\right]t = (\varepsilon_0\hbar/)t' = \omega_0 t'$, is dominated by the time at both inert frames and the time axis direction is always at the propagating direction, which is the optic axis of the electron microscope. Therefore, the front of the velocity field of the ensemble of the flying corpuscle electrons will dominate the phase variation of the component corpuscle electrons in the ensemble beam. That will result the individual corpuscle electron pulse rectifies its propagating orientation or frequency variation, coupling with each other in the spatial and/or temporal domain, which modifies the spatial and temporal distribution of the density of composited corpuscle electron pulses in the pulse ensemble beam due to the Talbot self-image. This spatial and temporal reconfiguration of the ensembles may emanate from the individual interaction between the corpuscle electron and the medium of the sample at a definite location and time. However, an observer using a microscope with an objective lens can only obtain an image of the density distribution of divergent electrons that enter the lens. This means that the information of the reconfiguration of the density of the corpuscle electron pulses ensemble is captured, but not the image of the medium of the sample. The near-field image is the spatial Talbot self-image of the corpuscle electron ensembles, and the far-field image, located at the back focal plane of the objective lens, is the temporal Talbot image, which exhibits the angular dispersion of the corpuscle electron ensembles as shown in Figure 3.11.

In the "corpuscle electron beam," there is a transversal periodicity L(x,y) which is related to spatial frequency $k_L = 2\pi/L(x,y)$ and a delayed time interval δt_0 which is related to frequency $\Omega = 1/\delta t_0$.

For TEM, the flying corpuscle electrons have a variable slanting angle to the optic axis or z-axis. This means the spectral condition is variable as $\Omega = (k_z - k_0)\mathrm{ctan}\theta$, which implies that the frequency Ω varies with the slanting angle θ and represents a plane in $(k_x, k_z, \omega/c)$-space that makes an angle (the spectral tilt angle) with respect to the k_z axis. The constraint $(k_x)^2 + (k_y)^2 + (k_z)^2 = (\omega/c)^2$ (which means the dynamic and kinetic energy is balanced in the corpuscle electron) imposes a relationship between the spatial and temporal frequencies in the paraxial region as $\Omega/c\,(1 - \cot\theta) = (k_x)^2 + (k_y)^2/2k_0$ resulting in a propagation-invariant envelope that is diffraction-free and dispersion-free in a vacuum:

$$\psi(x,z,t) = \int dk_x \tilde{E}(k_x,k_y)\exp\left[i(k_x x + k_y y)\right]\exp\{-i\Omega[t-(z/\mathrm{ctan}\theta)]\}$$
$$= \psi\{x,y,t-(z/\mathrm{ctan}\theta)\}$$

The velocity of the flying corpuscle electron can be modified by tailoring the spatiotemporal frequencies to change θ. The time-averaged intensity

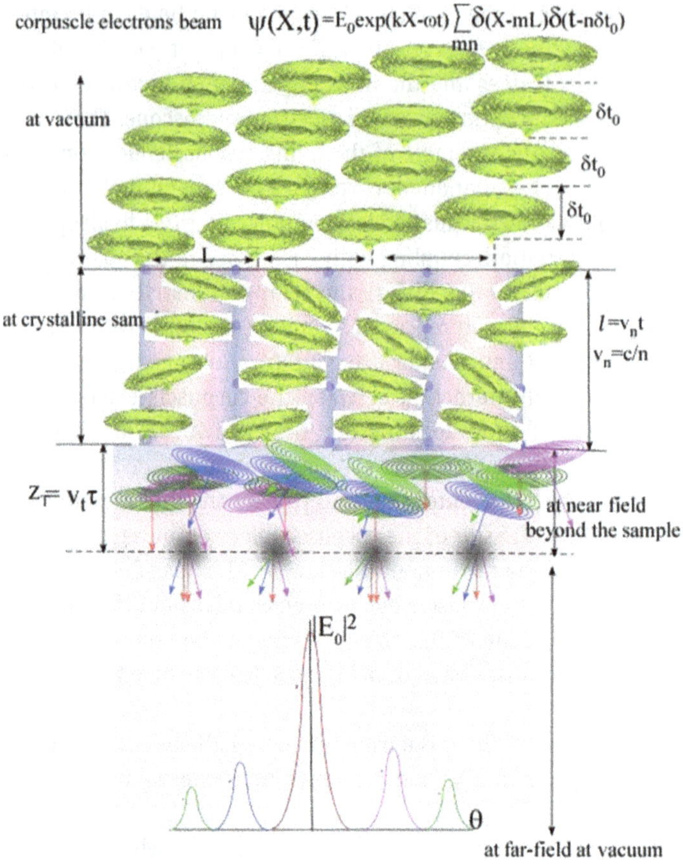

Figure 3.11: The corpuscle electron train, as "corpuscle electron bunching," travels through a crystalline sample, creating the Talbot self-image of the grating at near-field and the diffraction pattern at the far-field beyond the sample.

$$I(x,y,z) = \int dt|\psi(x,y,z,t)|^2 = \int dk_x\left|\tilde{E}(k_x,k_y)\right|^2$$
$$+\int dkx\tilde{E}(k_xk_y)\tilde{E}^*(-k_x,-k_y)\exp\left[2(k_xx+k_yy)\right]$$

(As it is recorded by a slow detector (CCD camera)) is independent of z and only related to the transversal spatial frequencies. For TEM, the recorded image and diffraction pattern always show the time-averaged intensity of the flying corpuscle electron ensemble. The intensity recorded shows the variation of the electromagnetic energy density of the corpuscle electrons with transversal spatial frequencies without temporal frequency.

If the wave function could be separated into spatial and temporal components at $z = 0$ as $\tilde{E}(k_x,\Omega) \approx \tilde{E}(k_x)\tilde{E}(\Omega)$ and $k_z - k_0 \approx \Omega/c - [(k_x)^2 + (k_y)^2]/2k_0$; $1/\tilde{v} = dk_z/d\Omega\,|_{\Omega=0}$ and $k_2 = d^2k_z/d^2\Omega\,|_{\Omega=0}$, then we can have the envelope function as

$$\psi(x,y,z,t) = \int dk_x \tilde{E}(k_x,k_y)\exp[i(k_x x + k_y y)]\exp\left[-i\left((k_x)^2 + (k_y)^2/2k_0\right)z\right]$$

$$\int d\Omega \tilde{E}_t(\Omega)\exp\{-i\Omega[t-(z/v_t)]\}\exp\{i(1/2)k_2\Omega^2 z\}$$

$$= \psi_x(x,y,z)\psi_t(t - z/v_t) \qquad \text{where } v_t = c\tan\theta$$

The $\psi_x(x,y,z)$ is the spatial envelope function, and the $\psi_t(t\text{-}z/v_t)$ is the corpuscle electron pulse temporal envelope traveling at a group velocity v_t, which is the dynamic contours of the Poynting momentum flux with the speed of light c. Due to the separability of the spatio-temporal spectrum with respect to k and Ω, we may discretize the spatial spectrum along k_x while maintaining a continuous variable in k_y. As we did before, the spatial envelope function may be given as

$$\psi_x(x,z) = \sum_{n=-\infty}^{\infty} \tilde{E}_n \exp(i2\pi nx/L)\exp(-i2\pi n^2(z/z_T))$$

$$\psi(x, mz_T, t) = \psi_x(x,0)\psi_t(t - mz_T/c)$$

That is, the transversal periodic spatial pattern undergoes revivals at the Talbot planes, accompanied by a temporal envelope propagating at a group velocity v_t independently of the spatial evolution of the field. In other words, the corpuscle electron pulses have the rotated circulating electromagnetic energy flux as a periodic energy oscillator at spatial element planes, accompanied by the momentum flux (Poynting vector) swift vorticity with light velocity c in the corpuscle electron frame. The scale of the temporal period may be seen as z_T/c. Because of the corpuscle electron, the spatial and temporal frequencies are coupled by energy conservation in the corpuscle frame, discretizing the spatial frequencies k_x entails that the temporal frequencies are also discretized as $\Omega \to n^2\Omega_L$, where $\Omega_L = (2\pi/z_T)[c/(1-\cot\theta)]$ corresponding to each spatial location in the corpuscle electron frame. In the corpuscle electron frame, the electromagnetic fields are perfectly revived at the Talbot planes, which is $x = mz_T$ with velocity $c = mz_T/t$, as $\psi(x,mz_T, mz_T/c) = \psi(0,0,0)$. The envelope observed in time at the Talbot planes ($ct = mz_T + c\tau$) may be expressed as

$$\psi(x, mz_T, (mz_T/c)+\tau) = \sum_{n=-\infty}^{\infty} \tilde{E}_n \exp(i2\pi nx/L)\exp\{-i2\pi n^2[(1/(1-\cot\theta))c\tau/z_T]\}$$

$$= \psi(x,0,\tau)$$

The time-resolved profile at the Talbot planes in the flying corpuscle electron frame corresponds to the profile at the initial plane $z = 0$. The axial coordinate z is replaced by $z = c\tau/(1-\cot\theta)$ and the profile observed in space along the axis in the traditional

Talbot configuration is observed in the time domain at fixed axial planes. The scale of τ is $\tau = z(1-\cot\theta)/c$ which closely relates to the velocity direction (or angle θ) of the corpuscle electron corresponding to the optical axis in the frame of the transmission electron microscope. That is the propagation of the de Broglie electron wave.

The time-average intensity of $I(x,z) = \int dt|\psi(x,z,t)|^2$ of the corpuscle electron can be expressed as

$$I(x,z) = \sum_{n=-\infty}^{\infty}\left|\tilde{E}_n\right|^2 + \sum_{n=-\infty}^{\infty}\tilde{E}_n\tilde{E}_{-n}{}^{*}\exp(i4\pi nx/L + \varphi_n - \varphi_{-n})$$

where $\tilde{E}_n = |\tilde{E}_n|\exp(i\varphi_n)$ is the discretized spectral field amplitude at $k_x = n(2\pi/L)$. The absence of any dependence on z means the underlying complex dynamics associated with time diffraction, which is veiled in physical space, and the transverse spatial period observed in physical space is L/2 rather than L. These spatial and temporal revivals of the electromagnetic energy density of the circulating relative vorticity field of the photon reveal the duality of the dynamics of vortex motion, which is the circulation and spatial translation. It may be more distinct that corpuscle electron current is the electromagnetic energy density annihilating and reviving in spatial and temporal domains such as the wave and particle duality of quantum particles in the laboratory frame.

3.5.2.5 Flowing corpuscle electrons and electron beam

We discussed the so-called de Broglie wave of an electron as electromagnetic pulses, having the spatio-temporal dynamics of the vorticity electromagnetic field of a photon and holding constant energy composed of kinetic and oscillating components. Its pulsation, as a function of the total energy $\mathcal{E}$ of the corpuscle electron, is given by $\omega = 2\varepsilon/\hbar$ The dynamic component energy varied with spatial and temporal expressions:

$$\omega = \sum_n 2\varepsilon_n/\hbar = \sum_n \omega_n = \sum_n ck_n = (\varepsilon_0\mu_0)^{-1/2}k_n$$

The pulsation of a corpuscle electron may contain many electromagnetic wave components with different frequencies. In other words, if a corpuscle electron is a pulse with an average of the frequencies, then the pulse train with a small separated time interval may be called a pulse beam having a mean frequency with small variation, or a coherent electron beam. If a certain number of corpuscle electrons with a separated time interval form a group and consecutive groups are separated by a larger time interval than the interval within the group, then it may be called corpuscle electron bunching. The corpuscle electron bunching may be separated by different time intervals, and the mean frequency of the pulses may also be different. If the corpuscle electrons have been separated by a huge distance, the corpuscle electron flow may be called single corpuscle electron flowing.

For electron microscopy, the corpuscle electron flow generally is a corpuscle electron beam or corpuscle electron pulse train as a consecutive flowing pulse train having a mean frequency with a small deviation, which determines the beam coherence as mentioned before. If it is a coherent beam, the behavior of the beam is the same as a single corpuscle electron pulse, but modified by the spatio-temporal dynamic behavior. In the laboratory frame, the flying corpuscle electron is a tiny electromagnetic energetic quantum particle, and its flying path is usually called a ray, and its internal dynamic behavior is usually expressed as a mechanical action function $\alpha(x,t) = S(x,t)/h$, which can be used in the Schrödinger and Hamilton-Jacobi equations at the microscope frame. Therefore, the phenomenon observed in an electron microscope should synthesize the physical process of the ray and the dynamic electromagnetic pulse train or bunching, which contain the single corpuscle electron and corpuscle electron pulse train as a beam. In the previous section, we discussed the behavior of a single corpuscle flying in free space. But now we discuss the corpuscle pulses beam interacting with the sample.

3.5.2.5.1 de Broglie electron wavefront and the corpuscle electrons pulse front

The de Broglie electron wavefront is the spatial geometric profile of the same phase at simultaneous times, expressing the spatial distribution of the same kinetic states at simultaneity in Minkowski space. This wavefront is dominated by the spatial frequencies $\mathbf{k}(x, y, z, t)$ or $\mathbf{k}(r, z, \omega)$ because the spatial frequency $\mathbf{k}$ is always perpendicular to the wavefront and determined by the frequency at the time elapsed and its track location. However, the front of a pulse is the spatial location of the earliest arrived electromagnetic energy flow with frequency ω_i. It is not simultaneous but it is the origin of an electromagnetic wave arriving in a pulse. The spatial contours of a pulse could be a step function or a Gaussian function. For a corpuscle electron pulse beam, each single pulse is almost the same, and the separated time intervals of consecutive pulses are very small relative to the beam current (for example, $\delta\tau = 10^{-9}$ s if the electron current is 100 pA). The pulse temporal duration in the propagation direction is related to the de Broglie wave frequency along this direction $\tau = 1/\omega_{\parallel}$. The spatial front of the corpuscle electron pulse beam is the assembly of simultaneously arrived corpuscle electrons at the same geometric location. If it is a cross-plane section of the beam, then the front of the pulse beam is a plane. A curve (convergent or divergent) or spiral cone (vortex or cone) is also possible, depending on the distribution of spatial arrival tracks coupled with the temporal delay characteristics of the pulse assembly. In other words, this induces an additional phase difference for the consecutive arrived corpuscle electrons, corresponding to the corpuscle electron arriving spatial location and relative temporal delay.

The electric field $E(x, y, t = z/v_t)$ of a flying corpuscle electron pulse is the spatial display of the dynamic process of the vorticity field of a rotating photon with energy $\mathcal{E} = h\omega$, which is $E(x, y, t = z/v_t) = [\mathcal{E}(x, y, t)]^{1/2} \exp[i\varphi(x, y, t)]$ expressed in the spatial and time domain or $\tilde{E}(x, y, \omega) = [\mathcal{E}^*(x, y, \omega)]^{1/2} \exp[i\phi(x, y, \omega)]$ in the spatial and frequency domain. For TEM, the spatial and time domain (or near-field) is located at or near the sample plane, and the spatial frequency domain locates the back focal plane (or far-field) of the objective lens. The intensity distribution of electric fields of the corpuscle electron pulse beam in the near-field is located near the object plane of the objective lens, which is recorded at the image plane of the objective lens, and the intensity distribution of electromagnetic energy is recorded at the back focal plane of the objective lens, which is the far-field. The spatial-spectral phase $\phi(x, y, \omega)$ is more important than the spatial-time phase $\varphi(x, y, t)$ of a corpuscle electron due to the complexity of understanding and measuring the spatio-temporal coupling of the corpuscle electron pulses induced by the interaction between the corpuscle electron and the sample during the pulse beam's traveling through the sample. Either the tested value at a location x_0, y_0, or a spatial average of this function over x, y, which is provided by a usual temporal measurement device (such as a CCD device), is closely related to the simple spectral phase $\phi(x, y, \omega)$. Measuring spatial-spectral intensity (or amplitude) may be simple, but measuring both intensity and phase is more challenging for electron microscopy.

For electron microscopy, a single flying corpuscle electron in free space or vacuum may be seen as confined vortex electromagnetic pulsated fields, showing a declining electric field at the laboratory frame and the spatially declining electric field emanating from the rotating electromagnetic field of a photon with a super-high frequency of 1,020 Hz. Due to the dynamic process at the internal frame of the corpuscle electron occurring in a vacuum, the velocity of the electromagnetic energy flux is homogeneous as light speed c, and the velocity of the corpuscle electron is $v_t = c\tan\theta$ $(\theta = v_t/c) = v_{\mathcal{E}} = \langle \mathbf{S}_k \rangle_{t,s} / \langle \mathcal{E}_k \rangle_{t,s}$, where $= \langle \mathbf{S}_k \rangle_{t,s}$, $\langle \mathcal{E}_k \rangle_{t,s}$ are the time and space average of the Poynting vector and energy as wave vector functions, as mentioned previously. Each pulse in the corpuscle electron pulse beam is almost similar if it is a coherent beam. However, if the corpuscle electron pulse enters a sample in which there exists a particular electromagnetic field distribution emanating from the structure and composition of the sample, which has a periodic or non-periodic distribution of permittivity ε or μ, then the dynamic process of the internal rotating vortex electromagnetic field of the photon should be modified $c = (\varepsilon\mu)^{-1/2}$ at that location. This is $c = (\varepsilon_0\mu_0)^{-1/2}$ based on the energy conservation law mentioned before:

$$\Delta k_t = \Lambda_t/hc \text{ or } \Delta\omega_t = \Lambda_t/h \; \Lambda_t(x', y', t = z/v_t) = \Lambda_t(r', \theta', t = z/v_t)$$

where x', y', r', θ', are the coordinates in the internal frame of the corpuscle electron; at this moment the corpuscle is overlapped with the sample in the laboratory frame, and the elapsed time t is in the laboratory frame. This $v_t = \Delta\omega_t/\Delta k_t = \langle \sum_{n(k),i} |\Im_{n(k),i}|^2 (\omega_i/k_n)\breve{u}_t \rangle$ is the group velocity or corpuscle electron velocity. During this dynamic process, the electromagnetic energy flux with different frequencies should be flowing in different orientations, and frequencies coupled to virtual spatial permittivity frequencies indicate that the spatial-temporal is coupled with each other. The pulse front would be tilted, and simultaneously, the de Broglie wave is dispersed with spatial and temporal frequencies. If the spatio-temporal frequency modification with angular θ then it is angular dispersion and rectifies the orientation of de Broglie wave propagation, resulting in the pulse beam front with angular dispersion and the particular pulse front being tilted or curved. In the spatial and frequency domain, if the pulse beam phase φ(x, y, ω) may be expressed as $\phi(x, y, \omega) = \eta x(\omega - \omega_0)$where the η expresses the magnitude of the angular dispersion, this relation can describe a phase varying linearly in frequency and position with a slope $\partial\phi/\partial\omega$, which is the pulse front tilted. Therefore, the beam affected by the track of the propagating corpuscle electron in the sample may be separated by the tilted beam front based on the different frequency components evolving at different tracks of the corpuscle electron. The beam's spatio-spectral and spatial-temporal properties would be changed due to propagation in the sample. If the spatio-spectral phase can be expressed as $\phi(x, y, w) = \xi r^2(\omega - \omega_0)^2$, where ξ represents the magnitude of chromatic curvature or the pulse front curvature. A slope of linear spectral phase varies quadratically with track position, or a quadratic spatial phase (wavefront curvature) varies linearly with frequency. That means if the spatio-temporal coupling of individual corpuscle electron pulses suffers from a slope quadratically varying linear spectral phase with its position or frequency, then the corpuscle electron beam would be affected by chromatism propagation. That results in the different frequency components evolving differently in space and time. The beam, as an ensemble of corpuscle electrons, has additional indecipherable spatio-spectral and spatio-temporal properties changed on propagation. If the wave function of the corpuscle electron expresses the spatio-temporal nature of the dynamics of the circulating vortex photon in its own frame, then the ensembles of the flying corpuscle electron as a beam not only have individual characters but also have the veiling future of the corpuscle electron pulses beam. These veiling properties of the corpuscle electron pulse beam can be exhibited at the near-field beyond the sample.

If the beam has a phase as $\phi(x, y, \omega) = \eta x\Delta\omega_t = \eta x\Lambda_t/h$, it travels in the spatial spectral domain at near-field, then the pulse front would be tilted, and the different frequencies will have different wavefront tilts in the near-field. These pulse front-tilted components of the beam would propagate with the tilted angular out of these locations. The feature of the Λ_t/h will dominate the distribution of the tilted angular or the spatio-spectral and temporal-spectral character of the corpuscle electron pulse beam, which will dominate the varying best focus position in the transversal dimension at far-field, which we may call "transversal spatial chirps" but usually call the diffraction pattern for electron microscopy. This is the result of the spatio-temporal frequencies transformation of the corpuscle electron pulse beam during propagation. The temporal structure of the corpuscle electron beam in the far-field is also very different from that in the near-field. As time passes, the corpuscle electron pulses with curvature front will focus at a spatial distance, and after the focus, if it loses the quadratic phase, it no longer has any pulse front tilting at far-field. However, the pulse in the near-field has a longer local temporal duration corresponding to the global duration. That is why the diffraction pattern only contains the spatial information.

3.5.2.5.2 The corpuscle electron pulse beam as flying torus pulse beam

As discussed previously, the corpuscle electron pulse may be seen as flying electromagnetic torus or doughnut pulses in which the pulse front is the front of the contours of the corpuscle electron in the propagating direction. The components of the dynamic electromagnetic waves are the circulating electromagnetic Poynting vector flow in the corpuscle's internal spatio-temporal frame with the central frequency $\omega_0 = m_0c^2/h$ and wave vector $k_0 = \omega_0/c$, and the broadband $\Delta\omega = \omega_t - \omega_0 = (\omega_{||}\omega_{\perp})^{0.5} - \omega_0$. Electric field of propagation of a single pulse may be expressed as

$$\begin{aligned}E(t,z) &= \int E_0(k)\exp(i(\omega_0 t - k_0 z)\exp\left[i(v_g(k-k_0)t - (k\ \ k_0)z)\right]dk \\ &= \exp(\omega_0 t - k_0 z)\int E_0(k)\exp\left[i(v_g t - z)(k-k_0)\right]dk \\ &= \exp i(\omega_0 t - k_0 z)E_1(z - v_g t) \\ &= \exp[ik_0(z - vt)]E_1(z - v_g t)\end{aligned}$$

In this equation, the function $\exp[ik_0(z - vt)]$, represents field oscillation with the central frequency, ω_0, which is the Zitterbewegung frequency, and the wavefront velocity, $v_{ph} = v$, is $v = \omega_0/k_0 = c/n(\omega)$. The part of $E_1(z - v_g t)$, represents the propagation of the envelope without deformation, determined by the group velocity $v_g = (d\omega/dk)\omega_0$. The $v = \omega_0/k_0$ phase velocity, v_{ph}, in a vacuum is the speed of light, but in a medium, n varies with frequency, ω. The wavefront velocity depends on the permittivity, n, varying with the frequency of the local track of the path of the flying electron. This results $k = (\omega/c)n(\omega)$in the pulse wavefront rectifying the propagation direction, and the group velocity of the corpuscle electron pulse varies $v_g = c/[n + \omega(dn/d\omega)]$. The pulse Poynting vector may be expressed as $<S(z - v_g t)> = (n/2\eta_0)E_1(z - v_g t)^2\exp(-2wkz/c)$, where $c = 1/(\varepsilon_0\mu_0)^{1/2}$and $\eta_0 = (\mu_0/\varepsilon_0)$. The electromagnetic energy flow with time may be expressed as

$$\begin{aligned}-\partial<\varepsilon>/\partial t = \nabla\bullet<S> &= (n/2\eta_0)\partial\ E_1(z - v_g t)^2/\partial z\ \exp(-2wkz/c) \\ &= -\{(\omega nk/\eta_0 c)E_1\bullet E_1^* + (n/\eta_0 v_g)(\partial E_1/\partial t)\bullet E_1^*\}\exp(-2\omega kz/c)\end{aligned}$$

This equation indicates electromagnetic energy flow related to ω, n, (μ_0/ε_0), v_g, and k. For corpuscle electron pulses, the spectral components coming from two ends of the spectrum propagate with different velocities, $\Delta v_g = (dv_g/d\omega)\Delta\omega$, with GVD defined as $\vartheta = d^2k/d\omega^2 = d(v_g^{-1})/d\omega = -(1/v_g^2)dv_g/d\omega$. The spectral components of the electromagnetic energy flow emanate from the spatial confinement of Poynting momentum flux with continuous time flow as a Möbius loop on the torus surface. This dynamic motion creates broadband electromagnetic oscillations with super-high frequencies around the center of the torus (k_0) and lower frequencies near the outer boundary of the torus (k_{max}), as discussed in Chapter 1. The flying corpuscle electron, as a flying torus of electromagnetic pulses, in which the confined electromagnetic energy flow constructs the spectral components of the electromagnetic waves with broadband frequencies, has a temporal period corresponding to its traveling periodicity as the spectral band scale $\Delta\omega = 1/\tau_0$ at the corpuscle electron frame. At microscope frame timescale is $t = \tau[1 - (v_t/c)^2]^{1/2}$ and the relation between these frames is $ct = \gamma c\tau + \gamma(v_t/c)z'$ defined. For a corpuscle electron torus pulse, the electromagnetic field having the group velocity with the highest frequency first arrives at the pulse front, and the field having the group velocity with the lowest frequency would be at the tail of the pulse. The spectral broadness is dominated by the frequency difference between the fields arriving at the pulse front and tail. For a corpuscle electron torus pulse, the propagating velocity of the Poynting flux always has the speed of light, c, and its circulation frequencies are dominated by the radius (or reciprocal vector $k \sim 1/r$) from the average rotation center (or torus central line) to the particular location of the vorticity field of the photon, as shown in Figure 3.5. This results in the highest frequencies being concentrated around the central line of the torus with the largest wave vector k_{max} (or smallest spatial area) and the lowest frequencies being distributed at the smallest wave vector $k_{\perp} = (\omega_t - \omega_0)/c$ (or spreading around the transversal contours of the corpuscle electron). Due to the linear variation of time flow, the linear propagation direction is the time-flowing direction, and its perpendicular plane would be the spatial domain of the confined electromagnetic energy flow motion. Therefore, the spatial-temporal coupling of the confined dynamic electromagnetic energy flow of a photon would occur. Due to the S(2) × S(2) topology geometric character of the torus configuration, there is symmetry of the rotation axis, which is the propagation direction, for the corpuscle electron pulse. Therefore, the flying corpuscle electron torus pulse beam is a special flying doughnut of electromagnetic pulses with isodiffraction properties.

3.5.2.5.3 Near-field feature of the flying corpuscle electron torus pulse beyond the sample

As discussed previously, the corpuscle electron beam can be seen as a flying electromagnetic energy torus pulse beam, and each electromagnetic field torus pulse has the following characteristics: (1) the propagation direction (z-axis in TEM microscope) is the temporal axis, (2) the space domain, which is perpendicular to the propagation direction, is the place in which the confined dynamic rotation of a circulating electromagnetic energy flux of a vorticity field of a photon occurs. The central region of the corpuscle electron torus pulse has the highest frequencies of the electromagnetic field, and the frequencies gradually decrease from the center to the boundary of the corpuscle electron torus pulse, (3) the distribution of the electromagnetic field of the corpuscle electron torus pulse has axis symmetry with mean frequency ω_0 and wave vector k_0, resulting in the isodiffraction feature, and (4) the spatio-temporal character of the confined electromagnetic field of circulating vorticity fields of a photon endows the dynamic process with spatial-temporal coupling, also called spectral dispersion of space and time. That may be expressed as

$$\omega(k) = \omega_0 + (d\omega/dk)(k - k_0) + 1/2(d^2\omega/dk^2)(k - k_0)^2$$

$$k(\omega) = k_0 + (dk/d\omega)_0(\omega - \omega_0) + 1/2(d^2k/d\omega^2)(\omega - \omega_0)^2$$

It shows that the pulse phase $\phi(x, y, \omega) = \eta x\Delta\omega_t$, $\phi(x, y, \omega) = \xi r^2(\omega - \omega_0)^2$, which is related to the pulse front being tilted or curved by the linear or quadratic variation at the transient track. The pulse front rectified varies with time elapsed and locus that is related to the local electromagnetic field of the medium. Therefore, pulse front variations occur at transient and locus, which means it occurs at the near-field of the pulse beam. That tells us the locus electromagnetic field at the medium (e.g., at an atom position) induces the transient frequency variation, which is encoded on the pulse front rectified at the near-field of the pulse train. At the far-field, this encoded information would be lost.

The quadratical spatio-temporal coupling induced different frequencies to have different wavefront curvatures in the near-field. But in the far-field, these different curvature waves must have a varying best focus position along the longitudinal dimension. At a single longitudinal position, this manifests as a varying beam size according to frequency and a spatio-spectral phase that represents the Gouy phase for each frequency. The pulses in time at focus have a more complex amplitude profile, with a longer duration on axis and arrival time that varies with the radial coordinate. The pulse front curved beam is significantly chirped everywhere in space, which gives the particular electromagnetic nature of a track locus. This would not change much the spatio-spectral picture because this simply corresponds to the addition of a spatially homogeneous spectral phase, but it would drastically change the picture in time. For the near-field, the spatio-spectral phase is quadratic in frequency, $\phi(x, y, \omega) = \xi r^2(\omega - \omega_0)^2$. It can be understood as a transversely varying linear temporal chirp, which in time corresponds to transversely varying pulse duration. This is also equivalent to the different colors having a wavefront tilt that varies quadratically with the frequency offset. In the focus, this frequency-varying tilt manifests as a quadratically varying best-focus position in the transverse dimension. In time at focus, the pulse amplitude is quite complex, but the temporal phase no longer exhibits any chirp because the chirps of different signs in the near-field average out at focus. In the near-field, the spatio-spectral phases can show the differential chirps which induce the pulse front tilt or curvature by the linear or quadratic frequency variation. The focus of the electromagnetic field of a pulse beam is emanated from pulse front tilting and curving due to the spatio-temporal coupling, but it is not by the lens. That is self-field focus.

Based on this understanding, the spatio-temporal coupling pulse train would have self-focusing, where a strong field of electromagnetic waves exists that may be seen as the spatio-temporal veiling Talbot effect. Along the pulse beam's propagating direction, the transversal dimension of the corpuscle electron torus pulse beam would have a variable transversal shape that may be called the shape of the wave function of the electron.

We notice that for TEM, the transversal intensity distribution of the corpuscle electron beam exhibits the spatial characteristics of the internal electromagnetic energy flux in the corpuscle electron frame. The intensity variation along the propagating direction (or z-axis) renders the temporal focus of the dynamic process of electromagnetic energy flux of the corpuscle electron, which shows as an energetic quantum particle with super-high oscillation having energy $\mathcal{E} = \omega\hbar/2$, called Zitterbewegung. It should be indicated that this is the duality of wave and particle for quantum particles such as electrons.

As shown in Figure 3.11, when a corpuscle electron enters the sample medium, the Poynting momentum flow (or phase velocity and group velocity) of the components of the dynamic electromagnetic fields of the rotating vorticity field of the photon will be modified by the permittivity, ε, and permeability, μ. The frequency also varies along the track locus of its traveling path with the time elapsed. The spatial-spectral phase variation accumulates over time, determined by the thickness, ℓ, of the sample. When the corpuscle electron exits the sample, the spatial-spectral accumulated phase ends, and the spatio-temporal coupling of the dynamic process of a corpuscle electron torus pulse in a vacuum is revived, but it already has additional fixed phases that carry particular spatio-temporal field fingerprints including linear and quadratic components. In the near-field, located beyond the exited surface of the sample, the tilted and curved pulse front would cross or focus, resulting in the spatial and temporal Talbot phenomenon beyond the exited surface of the sample, as shown in Figure 3.11. As mentioned before, the focusing points of different frequencies form an ensemble region with a high intensity of electromagnetic fields, which is the result of the spatio-temporal coupling of the confined electromagnetic field dynamic flux within the internal frame of the corpuscle electron. This can also be seen as a spatio-temporal Talbot phenomenon.

Since the value of the additional spatial-spectral phase accumulated relates to the sample thickness $\ell = v_g t$, the intensity distribution beyond the exit surface of the sample would vary with the corresponding thickness $\ell(x,y)$ that had been observed in TEM microscopy.

3.5.2.5.4 Corpuscle electron pulse beam travels at far-field

Maxwell's equations in the medium may be expressed as

$$\nabla \times \mathrm{H} = -i\omega\,\mathbf{D} \text{ and } \nabla \times \mathbf{E} = i\omega\,\mu_0 \mathbf{H},$$

Together with the constitutive relation $\mathbf{D} = \varepsilon_0 \varepsilon\,\mathbf{E}$, the spatial Fourier transform of the magnetic field is characterized by amplitude peaks $h_{n(k)}H_0$ located at the wave vectors $k_n = k + nK$, where $K = 2\pi/a$. The electric field $E_k(x) = E_k(x)\breve{e}_y$ and the electric flux density $D_k(x) = D_k(x)\breve{e}_y$ are unit vectors in the y direction. K is the reciprocal vector in the sample and can also be expanded as a Fourier series:

$$E_k(x) = \sum_n E_{n(k)} \exp(ik_n x) \quad D_k(x) = \sum_n D_{n(k)} \exp(ik_n x)$$

The Fourier coefficients of $E_{n(k)}$ and $D_{n(k)}$ may be expressed as

$$E_{n(k)} = e_{n(k)} \mu_0 c H_0 \quad D_{n(k)} = d_{n(k)} H_0 / c$$

The $e_{n(k)}$ and $d_{n(k)}$ are dimensionless coefficients. If the Fourier expansion of the permittivity ε(x) could be

$$1/\varepsilon_{(x)} = \sum_n k_n \exp(inKx)$$

Then, we may yield the following relations between the Fourier coefficients $h_{n(k)}$, $_{(k)}$, and $d_{n(k)}$:

$$d_n(k) = h_{n(k)}(k_n c)/\omega \quad h_{n(k)} = e_{n(k))}(k_n c)/\omega \quad e_{n(k)} = \sum_{n'} \kappa_{n-n'} d_{n'(k)}$$

For each integer, n should be held:

$$\sum_{n'} \kappa_{n-n'} k_n k_{n'} h_{n'(k)} = (\omega/c)^2 h_{n(k)}$$

The spatial-temporal average of the electromagnetic field energy $<\mathcal{E}_k>_{t,s,}$ can also be decomposed into fractional energy densities $\varepsilon_{n(k)} = 1/2 \mu_0 |h_{n(k)}|^2 H_0^2$ corresponding to the energy density of an electromagnetic plane wave with the magnetic field amplitude $h_{n(k)} H_0$. Similarly, the time–space average Poynting vector $< \mathbf{S}_k >_{t,s}$:

$$< S_k >_{t,s} = < \mathrm{Re}\,(ExH^*)/2 >_s = \sum_n \ (1/2\ ) \mu_0 c\, |h_{n(k)}|^2 H_0^2 (\omega / k_n c) \mathbf{x_0}$$

The expanding Poynting vectors are $<S_{n(k)}> = (1/2) \mu_0 c\, |h_{n(k)}|^2 H_0^2 (\omega / k_n c) \mathbf{x}_0 (\mathbf{x}_0)$ unit vectors of the propagating direction corresponding to the Poynting vector of an expanded electromagnetic plane wave with the wave vector k_n and the magnetic field amplitude $h_{n(k)} H_0$. When the *n*th component is considered individually, it always corresponds to the case of an electromagnetic plane wave with the wave vector k_n and the magnetic field amplitude $h_{n(k)} H_0$. This shows that an electromagnetic wave can be decomposed into a series of component electromagnetic plane waves. The expanding *n*th plane wave is characterized by the magnetic field amplitude $h_{n(k)} H_0$ and the wave vector k_n. (In the previous chapter, we discussed spatial-temporal energy exchange between vortex and translating electromagnetic fields.) Its contribution to the corpuscle's field is given by $|h_{n(k)}|^2$, which is the ratio between the energy carried by this plane wave and the total energy carried by a corpuscle electron. The *n*th plane wave is not an electromagnetic plane wave because it does not individually satisfy Maxwell's equations, unlike the general electromagnetic wave, and it is not an eigenvector for the eigenvalue ω. The group velocity $\mathbf{v_g}$, which is equal to the energy flow velocity $\mathbf{v_e}$ in periodic media:

$$\mathbf{v_g} = \mathbf{v_e} = \langle \mathbf{S}_k \rangle_{t,s} / \langle \varepsilon_k \rangle_{t,s} = \sum_n |h_{n(k)}|^2 (\omega/k_n)\mathbf{x}_0;\ v_n = (\omega/k_n)\mathbf{x}_0$$

The group velocity of a corpuscle electron pulse is the group velocity $v_n = \omega / k_n$ of the expanding plane waves weighted by their energetic contribution $|h_{n(k)}|^2$. The solution to Maxwell's equations at energy $h\omega_n$ in the sample is an electromagnetic plane wave with a wave vector. $k_n = \sqrt{\varepsilon}\omega_n/c = \omega_n/v_{g(n)}$. This plane wave can be regarded as a particular component wave for which $h_{n(k)}$ is equal to 1 for a unique integer n^*, and 0 otherwise. The electromagnetic wave decomposition is totally dominated by the single plane-wave solution of Maxwell's equations. The index of these component waves is labeled n^* and $n^* = 1$ for the component waves within the first spatial frequency band, $n^* = 2$ for the second band, and so forth.

If we think of the two-dimensional case, the Fourier expansion of $1/\varepsilon(x,y)$ would be as

$$1/\varepsilon(x,y) = \sum_{n,m} \kappa_{n,m} \exp(iG_{n,m} \bullet r); H_k(r) = \sum_{n,m} |h_{n,m(k)}| H_0 \exp[i(k + G_{n,m}) \bullet r]\check{\mathbf{Z}}$$

$$E_{n,m(k)}(r) = \Sigma_{n,m} e_{n,m(k)} \mu_0 c H_0 \exp[i(k + G_{n,m}) \bullet r]\check{\mathbf{Z}}$$

$$D_k(r) = \sum_{n,m} d_{n,m(k)} H_0/c \exp[i(k + G_{n,m}) \bullet r]\check{\mathbf{Z}}$$

$$d_{n,m(k)} = [(k_{n,m}c)/\omega]x\, h_{n,m(k)}\quad h_{n,m(k)} = [(k_{n,m}c)/\omega]x\, e_{n,m(k)}$$

$$e_{n,m(k)} = \sum_{n',m'} \kappa_{n-n',m-m'} d_{n',m'(k)})$$

$$\sum_{n',m'} \kappa_{n-n',m-m'} |k_{n,m}||k_{n',m'}| \left(e_{n',m'(k)}/|k_{n,m}|\right) = (\omega/c)^2 \left(e_{n',m'(k)}/|k_{n,m}|\right)$$

As discussed before, the group velocity is given by the vectorial sum of the phase velocities $\omega\mathbf{k}_{n,m}/\mathbf{k}^2_{n,m}$ weighted by the energetic contribution $|h_{n,m(k)}|^2$ of the corresponding plane waves. Each term can have a physical interpretation that provides an intuitive understanding of the direction of the group velocity.

For TEM, it is valuable to emphasize that the wave vector $\mathbf{k}_{n,m}$ is the Poynting component of the electromagnetic fields in the corpuscle electron frame. As mentioned before, it has the spatio-temporal coupling of the transient wave vectors and frequencies. The temporal varying axis along $k_{||}$ and spatial profile variation in transversal $k_{\perp}$ and the divergent angle α of the propagating corpuscle electron pulse are $\alpha = k_{\perp}/k_{||}$ important. The waves traveling transversally in the sample are the wave $k_{\perp}$; therefore, this $\mathbf{k}_{||n,m} = \mathbf{k}_{n,m} = \mathbf{k} + \mathbf{G}_{n,m}$ would be the Bragg condition. The group velocities of the de Broglie longitudinal components in periodic atom configuration can be given by the gradient vectors of the isofrequency surface of the corresponding Brillouin zone. The equation $\mathbf{k}_{||n,m} = \mathbf{k}_{n,m} = \mathbf{k} + \mathbf{G}_{n,m}$ gives the series of spatial loci of transmitting group velocity, at which the diffracted component electromagnetic waves are traveling in the direction of $\mathbf{k} + \mathbf{G}_{n,m}$ and have the energy fraction $(k_{n,m}c)/\omega$ with $\mathbf{k}_{||n,m} = \mathbf{k} + \mathbf{G}_{n,m} = 2n\pi$.

The isofrequency surface at the first Brillouin zone may be different based on the particular atom character and symmetric configuration. The gradients of the isofrequency surface of the corpuscle electron torus pulse match with the iso-spatial frequency surface, which holds the momentum and energy conservation in the crystal sample. The total of the fractional electromagnetic energy, $|h_{n,m(k)}|^2 H_0^2$, of diffracted pulse beams should equal the energy of the corpuscle electron, $\hbar\omega$, which is the average of variable frequencies $\omega = (\omega_{||}\omega_{\perp})^{1/2}$. It is clear that the ratio between the energy of the (n,m) plane wave and the total energy carried by a corpuscle electron is determined by the quantity $|h_{n,m(k)}|^2$ and the wave vector $\mathbf{k}_{n,m}$ is the electromagnetic energy flowing direction. The group velocity is given by the vectorial sum of the phase velocities $\omega\mathbf{k}_{n,m}/k_{nm}^2$ weighted by the energetic contributions $|h_{n,m(k)}|^2$ of the corresponding plane waves. However, while corpuscle electron pulses propagate in the sample instead of a vacuum, the group velocity of corresponding component plane waves would be manipulated by the spatial electromagnetic field of the sample, which varies with the time elapsed due to $\Delta k_t = \Lambda_t/h$ or $\Delta\omega_t = \Lambda_t/h$, $(\Lambda_t(x', y', t = z/v_t) = \Lambda_t(r', \theta', t = z/v_t)$as discussed previously. For a corpuscle electron having energy conservation, this means $\omega^2 = \omega_0^2 + (ck)^2$that it endows the spatial-temporal coupling. If the $\omega_0 = ck_0$ is the original frequency and wave vector at rest condition of the corpuscle electron in a vacuum, and we already know that $\omega^2 = \omega_{||}\omega_{\perp} = c^2k_{\perp}||k_{\perp}$,then $k_{||}k_{\perp} = k_0^2 + k^2$, where k is the transient wave vector and $k_{||}$,$k_{\perp}$ are the longitudinal propagating wave vector and transversal wave vector. The de Broglie wave vector is the $k_{||}$ along the propagation direction, which is the time axis. The transversal wave vector $k_{\perp}$ is the spatially transmitted wave vector and $k^2 = k_{||}^2 + k_{\perp}^2$.

Therefore, we may think the transmitting diffraction beams are transmitted components of electromagnetic waves with energy $|h_{n,m(k)}|^2 (\omega/c)^2$, but as previously indicated, $\omega^2 = \omega_{||}\omega_{\perp} = c^2k_{||}k_{\perp}$ the energy of a diffracted wave beam would vary, $\mathbf{k}_{n,m} = \mathbf{k} \pm \mathbf{G}_{n,m(k)}$ and the energy distribution inside the transmitted original corpuscle electron pulse beam would have the fingerprint of the energy deficits (observed as Kikuchi bands) induced by the electric potential of the sample.

This iso-frequency analysis demonstrates that the gradient vector of the isofrequency surface of the Brillouin zone induces the group velocity tilting, resulting in diffraction at the far-field. Analysis of the diffraction divergent angle would obtain the distribution of the group velocity related to the spatial-temporal frequency of the Brillouin zone of the crystalline sample.

The space-time diffraction process may be simply interpreted as follows. The incident electromagnetic wave in the corpuscle electron frame is refracted into the sample medium at $z = 0$, while the circulating vortex field of the photon generates an infinite set of time harmonics inside the sample medium, with frequencies $\omega_n = \omega_0 + n\Omega$ corresponding to the wave vectors $k'_n = k'_{||} + n\Omega/v'_r$ (the v'_r is the tangent velocity of the rotation). The refracted space-time plane waves in the sample, which are also in the corpuscle frame, are diffracted into an infinite set of plane waves traveling toward the $z = \ell$ boundary. The phases of the space-time harmonic waves inside the sample accumulate over time as the longitudinal wave passes through the sample. Then, while the space-time harmonic waves escape the interface between the sample and vacuum, propagating into the vacuum again, these space-time harmonic waves in the corpuscle frame have undergone phase matching during the escaped propagation and the evanescent waves in the region near the interface. The x components of the wave vectors of the mth mode in vacuum and the x component of the wave vector of the mth space-time harmonic field accumulated in the sample must be the same. To determine the spatial and temporal frequencies of the diffracted orders, we consider the momentum conservation law, i.e.

$$G_{x;di-} = k_{x;inc} + mG; \quad \text{or} \quad k''_{x;mn} = k_{x;mn} = k_{||x;i} + mG$$

and the law of energy conservation, i.e.,

$$\omega_{diff} = \omega_{inc} + n\Omega; \text{or } \omega_n = \omega_0 + n\Omega$$

where $k_{x;diff}$ and $k_{x;inc}$ denote the x components of the wave vector of the diffracted and incident fields, respectively. ω_{diff} and ω_{inc} represent the temporal frequencies of the diffracted and incident fields, and G is the spatial frequency of the sample, respectively:

$$k''_{||} + n\Omega/v''_r \sin(\theta'' mn) = k_{||}(\theta_i) + mG$$

where $k_{||} = \omega_{||}/c$. seeking the angle θ''_{mn} of diffraction for the forward spatial-temporal diffracted orders in the region outside of the sample, i.e., the mth spatial and nth temporal harmonic, it yields

$$\sin(\theta''_{mn}) = (\sin(\theta_i) + mG/k_{||})/(1 + n\Omega/\omega_0)$$

The corresponding angle θ_{mn} of diffraction for the backward diffracted orders in vacuum would be

$$\sin(\theta_{mn}) = (\sin(\theta_i) + mG/k_{||})/(1 + n\Omega/\omega_0)$$

While the flying corpuscle electron pulse is perpendicularly entering the interface between the sample and vacuum, due to $\theta_i = 0$

$$\sin(\theta''_{mn}) = (mG/k_{||})/(1 + n\Omega/\omega_0)$$

Due to $1/(1 + n\Omega/\omega_0) \approx 1 - (n\Omega/\omega_0)$ then $\sin(\theta''_{mn}) \approx [(mG/k_{||}) - (n\Omega/\omega_0)(mG/k_{||})]$ then, if the $G/k_{||}$ is very small, i.e., the paraxial case, the transmitted longitudinal wave vector is larger than the spatial frequency. $G = 2\pi/a$, if we neglect the temporal effect, then $\sin\theta''_{mn} \approx \theta''_{mn} \approx G_m/k_{||}$, which is usually used for the Bragg angle. The harmonic oscillation frequency can manipulate the diffracted angle θ''_{mn} as $\theta'' \approx \theta''_{mn} - \Delta\, \Delta\theta''_{mn}$.

Therefore, analysis of the diffraction beam intensity distribution would obtain the atomic dynamic condition in the sample.

3.6 What is transmission electron microscopy?

Based on the previous section's discussion, albeit naïve, we may clear out the bewilderment regarding TEM.

Using TEM felicitously deciphers the clumsy duality of wave and particle of an electron statistically and dynamically as the first chapter mentioned. The intrinsic natures of the spatial and temporal characters of the wave and particle of an electron have been demonstrated in an electron microscope. It is clear, at least for me, that an electron is a physical entity of spatially confined electromagnetic energy flux emanated from dynamic circulating vortex electromagnetic fields of a photon with S(2) × S(2) topological geometry having 4π periodicity. The energy of a flying electron has two parts: (a) intrinsic circulating of vorticity field of a photon having temporal frequency as $2\omega_0 = m_0c^2/h$ called Zitterbewegung. This electromagnetic Poynting flux rotates around the center of radiated rotating moment having spatial scale $\lambda_{comp} = h/m_0c$ that is the static state of the electron at the laboratory frame; (b) for a flying electron, the frequency of the electron would be $\omega = \gamma\, m_0c^2/h$ and the de Broglie wave is the confined waves induced by the relative motion of the internal electromagnetic field flux in the frame of the corpuscle electron. These confined electromagnetic waves are the Doppler waves of the electromagnetic wave in the corpuscle electron frame, in which the flying direction of the corpuscle electron has the waves with blue shift but the wave with red shift on the perpendicular to the flying direction. These Doppler waves have the spatial-temporal coupling that results in the duality of spatial and temporal motion in which the spatial swirling and temporal translation.

For TEM, while the linear magnification of the electron microscope reaches $>\times 10^6$ meaning the dimensionless scale is smaller than $\times 10^{-6}$, the electron is a quantum particle and corpuscle with a dynamic electromagnetic field having $e/m = 1.76 \times 10^{11}$ C/kg. The contrast on the image observed at the display device may be understood as the energetic quantum particles' density spatial distribution, called mass-thickness contrast in the last century. However, when magnification is larger than $\times 10^6$ or $\times 10^6 - \times 10^8$ the electron would be the corpuscle electron with the confined dynamic electromagnetic vorticity field of a photon, and the antibunch characteristics of the electromagnetic field of the particle may be relaxed depending on the spatial separations and transient temporal interval between the aggregated corpuscle electrons. The electromagnetic field interaction would dominate the kinetic and dynamic process of motion in the spatial and temporal domain, resulting in the duality of circulating motion and linear motion or duality of space and time. At high magnification ($>10^7$), the visible linear scale at the microscope would be reduced to 10^{-11} m or 10^{-2} nm, and the timescale would be as small as $\delta t = \delta s/v_t$. If v_t is constant, δt may be at 10^{-11} s or more than nanoseconds. Therefore, at high magnification in TEM, the transversal kinetic and dynamic process of the corpuscle electron pulsed beam may be observed at the nanoscale of space (near-field or the objective plane of the lens) and time (far-field or back focus plane of the imaging lens) domain.

The corpuscle electron is an electromagnetic pulse. In the laboratory frame, the corpuscle electron has a spatial configuration (i.e., profile) or contour, in which the internal dynamic electromagnetic fluxes have the motion of rotating and circulating of the vortex electromagnetic field of the photon at the corpuscle electron frame that seems like a flying electromagnetic torus pulse with spatial-temporal coupling. The spatial surface of the contour may be seen as the pulse front, and the consecutively arriving electromagnetic energy flow with time delay would be the internal dynamic electromagnetic waves of the corpuscle electron. The frequency spectrum of the pulse is the broadband of the Doppler waves. As shown in Figure 3.2, the isofrequency surface of the spatial-temporal coupling demonstrates the highest frequencies around the center of the corpuscle pulse and gradually symmetrically decreasing from the center to the edge of the contour, which gives the isodiffraction. That is why a collimated electron with a hydrogen atom passes through the hydrogen atom without changing traveling direction or scattering. The corpuscle electron pulse front profile could have different geometric shapes such as a plane or curved surface emanated from the time distribution of the de Broglie phase wave arriving at the pulse front. The pulse phase gradient and curvature steer the tilting and focusing of the pulse front. The quadratic curvature front triggered by the transient field would form a focus point at the pulse near-field as the Talbot self-focusing effect with spatial-temporal coupling. The spatio-temporal coupling of the electromagnetic component's energy fluxes in the corpuscle electron frame with frequency band $\Delta\omega = \omega_{\parallel} - \omega_{\perp}$ and the pulse front shape variation triggered by the transient field at each track of the passed path. If this understanding is acceptable, the TEM is the quantum electron microscopy which contains (1) the duality of the particle and wave emanated from the duality of vorticity and linear motion, (2) space-time coupling or spatialtemporal separability and nonseparability, (3) analysis of electromagnetic field intensity distribution at the near-field of the corpuscle electron pulse train or beam after passing an object (atom, crystal, molecular, amorphous object), (4) spectrum analysis after passing an object containing analysis of the velocity of the quantum particles and frequency (spatial and temporal) analysis of electromagnetic Poynting momentum fluxes at the far-field of the corpuscle electron pulse beam, and (5) analysis, which has not developed yet, is the measurement of the quantum phase with time for the corpuscle electron pulse beam. The spatial periodicity of the object can be exhibited at near-field using the spatial-temporal coupling Talbot effect, but the character of the particular atom field encoded the pulse front shape which is steered by $\partial\phi/\partial s$ and $\partial^2\phi/\partial s^2$. If we wish to reveal this information, the space and timescales should be around 10^{-13} m and less than 10^{-10} s. In TEM, the magnification can be $\times 10^7$–10^8 which implies the space and timescale at this magnification could be smaller than $\times 10^{-8}$–10^{-9}. To steer the corpuscle electron pulse beam transversal profile to combine with an additional strong super-high frequency optical field may decipher the clumsy phase information.

We would like to emphasize that the spatial-temporal coupling Talbot pattern exhibits the spatial periodicity distribution of the intensity of the corpuscle electron pulse beam, which corresponds to the periodicity of the atom grating, but not the projection of the atom potential on the screen of the microscope or the interference pattern of the Bloch electron plane wave. Because the sample of the microscope is located at the front focal plane of the objective lens of the microscope, the electrons exiting the surface of the sample act as the objective plane of the objective lens. Therefore, the electromagnetic field intensity distribution of the corpuscle electron pulse beam at the near-field beyond the sample may be the self-imaging Talbot pattern of the beam, and the Talbot plane at $z_T = L^2/2\lambda_{dB}$ should be the objective plane, which is near the front focal plane of the objective lens. That is the reason for using a serious defocus method to get the high-resolution image corresponding to the preferred spatial characteristics of the crystal structure of your sample.

Based on previous discussions, each diffraction beam should have different energy, $|h_{n,m(k)}|^2(\omega_{n,m}/c)$, and a deflected angle $\Delta\theta$ related to the optical axis or incident direction, which is related to time delay. Using velocity analysis (energy loss spectrum) of different order diffraction spots may reveal the linear chirp of atoms in the sample, and using an illuminating beam with a large convergent angle produces the off-focusing shift along the optical axis. We may see the interference pattern between the nearest diffracted beams that may detect the phase difference of these corpuscle electron pulse beams.

I would like to indicate that the aberration-corrected objective lens is a great contribution to electron microscopy, but deciphering the information of interaction between the electron and matter contained in the corpuscle electrons on the image of the microscope is not the duty of the electron microscope. Understanding the electron as an illuminating source of the microscope or the electron beam may be more important as A. Howie indicated. However, the electron is the center of modern physics and the focal point of quantum mechanics. Therefore, electron microscopy would be a paradise for quantum physics players.

References

Tomilin, A.K. “The Potential-vortex theory of electromagnetic waves.” Journal of Electromagnetic Analysis and Applications. 5, 347–353, (2013).

Zdagkas, A., Papasimakis, N., Savinov, V. and Zheludev, N.I. “Building blocks for space-time non-separable pulses. ”arXiv:1912.09332v1 [physics. optics] (19 Dec 2019).

Zdagkas, A., Papasimakis, N., Savinov, V., Dennis, M.R. and Zheludev, N.I. "Singularities in the flying electromagnetic doughnuts." Nanophotonics. 8(8), 1379–1385, (2019).

Zdagkas, A., Papasimakis, N., Savinov, V. and Zheludev, N.I. "Space-time nonseparable pulses: Constructing iso-diffracting donut pulses from plane waves and single-cycle pulses." Physical Review A. 102, 063512, (2020).

Lohnann, A.W., Knuppertz, H. and Jahns, J. "Fractional Montgonery effct: A self-imaging phenomenon." Journal of Optical Society of America A. 22(8), 1501, (Aug 2005).

Byoung, S.H. "Deterministic control of photonic de Broglie waves using coherence optics." Scientific Reports. 10(12), 899, (30 July 2020).

McMorran, B.J. and Cronin, A.D. "An electron Talbot Interferometer." New Journal of Physics. 11, 033021, (2019).

Dolce, D. "Intrinsic periodicity: The forgotten lesson of quantum mechanics." J. Phys.: Conf. Ser. 442 (2013). Donatello.dolce@coepp.org.au.

Hestenes, D. "Quantum Mechanics of the electron particle-clock." arXiv:1910.10478v2 [physics. Gen-ph] (24 Jan 2020).

Hestenes, D. "Zitterbewegung structure in electrons and photons." arXiv:1910.11085v2 [physics. gen-ph] (24 Jan 2020).

Hestenes, D. "Deconstracting the electron clock." (July 2018). URL: http://geocalc.clas.asu.edu/.

Jones, E.R., Bach, R.A. and Batelaan, H. "Path integrals, matter waves, and the double slit." European Journal of Physics. 36, 065048, (20pp), (2015).

Logiurato, F., "Relativistic derivations of de Broglie and Planck-Einstein Equations." arXiv:1208.0119v1 [quant-ph] (1 Aug 2012).

Bartłomiej Kiałka, F. "Talbot-Lau Schemes and Bragg Diffraction: Theory and Applications in High-Mass Matter-Wave Interferometry." Thesis, Universtät Wien, 0807, (2021).

Maria Vanacore, G., Madan, I. and Carbone, F. "Spatio-temporal shaping of a free-electron wave function via coherent light–electron interaction." La Rivista Del Nuovo Cimento. 43, 567–597, (2020). https://doi.org/10.1007/s40766-020-00012-5.

Salas, J.A., Varga, K., Yan, J.-A., Kirk and Bevan, H. "Electron Talbot effect on graphene." Physical Review B. 93, 104305, (2016).

Louis Van Belle, J., "The electron as a harmonic electromagnetic oscillator." (1 June 2019).

Louis Van Belle, J. "Einstein's mass-energy equivalence relation: an explanation in terms of the Zitterbewegung." (24 Nov 2018).

Louis Van Belle, J. "De Broglie's matter-wave: concept and issues." (9 May 2020).

Macken, J.A. "Spacetime Based Foundation of Quantum Mechanics and General Relativity." Progress in Theoretical Chemistry and Physics. 29, Springer Switzerland, pp. 219-245, (2015).

Halli, L.A., Yessenovi, M., Ponomarenko, S.A. and Abouraddy, A.F. "The Space-Time Talbot effect." arXiv:2102.06769v1 [physics. Optics] (12 Feb 2021).

Halli, L.A., Yessenovi, M. and Abouraddy, A.F. "Space-time wave packets violate the universal relationship between angular dispersion and pulse-front tilt." arXiv:2101.07317v2[physics. Optics] (20 Jan 2021).

Halli, L.A., Ponomarenko, S.A. and Abouraddy, A.F. "Temporal Talbot effect in free space." arXiv:2103.12677v1 [physics. Optics] (23 Mar 2021).

Chapman, M.S., Ekstrom, C.R., Hammond, T.D., Schmiedmayer, J., Tannian, B.E., Wehinger, S. and Pritchard, D.E. "Near-field imaging of atom diffraction gratings: The atomic Talbot effect." Physical Reviews A. 51(1), 51, (Jan 1995).

Mir, M., Bhaduri, B., Wang, R., Zhu, R. and Popescui, G. "Quantitative Phase Imaging." Progess in Optics. 57, 133 (© 2012). Elsevier B.V.

Butto, N. "A new theory on electron wave-particle duality." Journal of High Energy Physics, Gravitation and Cosmology. 6, 567–578, (2020).

Vaudon, P. "The de Broglie wave in the solution of Dirac equation." (2020). Hal–02448288v6.

Peter, J.O. "Dispersive quantization- the talbot effect." (2014). http://www.math,umn.edu.
Tainta, S., Erro, M.J., Garde, M.J. and Muriel, M.A. "Temporal self-imaging effect for periodically modulated trains of pulses." Optics Express. 22(12), 15251, (16 June 2014).
Jolly, S.W., Gobert, O. and Quéré, F. "Spatio-temporal characterization of ultrashort laser beams: A tutorial." Journal of Optics. 22, 103501, (2020).
Shen, Y., Zdagkas, A., Papasimakis, N. and Zheludev, N.I. "Measures of space-time non-separability of electromagnetic pulses." arXiv:2006.12603v1 [physics. optics] (22 Jun 2020).
Shvyd'ko, Y. and Lindbergy, R. "Spatiotemporal Response of Crystals in X-ray Bragg Diffraction." arXiv:1207.3376v1 [physics. optics] (13 Jul 2012).

Chapter 4
Quantum electron microscopy

4.1 Understanding what is a corpuscle electron pulse beam and what is an electron beam

As discussed previously, an electron is an energetic particle and an assembly of tracks of a circulating vorticity photon with speed of light. In short, a quantum particle with duality of particles and waves or fields is called an electromagnetic field. In the laboratory frame, as electron microscope, the flying electron tracks form the paths of trajectories of the electrons near the optic axis of transmission electron microscope. As electrons have charge and if its flying velocity is v_t, the flying electrons create the electric current I_e (as mentioned $I_e \approx 10^2 \sim 10^3$ pA (picoampere)), accompanying with magnetic field, but two consecutive electrons cannot close each other due to electromagnetic field action, which is its anti-bunch property of charge particles in the laboratory frame. This property roots in the Maxwell equation, which is an electric charge current-induced magnetic field and two parallel flying charge-induced magnetic fields would have the repelling force. However, the electric charge of an electron is the result of the circulating vorticity electromagnetic field of dynamic moving photon at the electron's own internal frame. Therefore, any spatio-temporal points (x_i, y_i, z_i ($=v_t t_i$)) on the paths of the trajectory of a flying electron at the laboratory frame must simultaneously have the corresponding dynamic parameters corresponding to the relative spatial-temporal point (x',y',z'($=v_t't'$)) at the electron's own internal frame. In other words, kinetic parameters of flying electrons at the laboratory spatial time frame should connect to the dynamic parameters of the circulating vorticity photon in the electron's own internal spatial time frame. This correlation may be expressed as the action function of a free relativistic particle such as

$$S = mc^2 \int_{t_1}^{t_2} \left(1 - v_t^2/c^2\right)^{0.5} dt \quad \text{or} \quad \delta S = mc^2 \delta t \left(1 - v_t^2/c^2\right)^{0.5} \approx \varepsilon_t \delta t - p_n \delta x_n$$

In the laboratory frame, the electron is a linearly flying charge particle, for which its mass $m = \gamma m_0$ emanated from the center of the rotating Poynting energy flux and its charge, q, originated from the vorticity field of the circulating photon. The ratio of charge and mass, $e/m = 1.8 \times 10^{11}$ C/g, is the fundamental physical parameter of an electron. The kinetic state of a flying electron may be described as $\psi = E_0 \exp(-iS) = E_0 \exp\{-i[\varepsilon_t \delta t - p_n \delta x_n]\}$, which is usually used as the wave function of de Broglie's electron wave, and $\varepsilon_t = (\gamma m_0 c^2 + m v_t^2/2) = |\omega| h$ and $\mathbf{p}_n = \mathbf{k}_n h$. Therefore, the action function is a bridge to correlate the laboratory frame with the internal frame of an electron. It is valuable to emphasize that the action function (or phase) is the time difference between the laboratory frame and the moving corpuscle electron frame,

https://doi.org/10.1515/9783111449333-004

which implies that the dynamic motion process may be similar but the evolution time difference would dominate the particular physical motion process in different frames. Therefore, the spatial-temporal process is the key parameter.

The rotating or circulating motion has spatio-temporal periodicity of kinetic states, and the linear motion has continuous flowing time. The flying corpuscle electron has this motion duality. These motion processes have to be understood as a space-time coupling process. The concept of space-time coupling is an important physical concept that has fused the three dimensions of space and the one dimension of time into a single four-dimensional continuum called Minkowski's space-time. The causality of any motion events has been involved, and the relationship between the events occurred at any spatial location and its corresponding time has been coupled. For example, the wavefront is the assembly of same dynamic states of the harmonic oscillator at space with simultaneity. Wavefronts of the de Broglie wave are the assembly of dynamic simultaneity phase states of circulating vorticity photon at its internal frame space, but the spatial contour of this internal frame is the spatial profile of the electron that visualized at the laboratory frame. At the internal frame, the electric field originated from the vorticity magnetic field, which varies with transient time, but at the laboratory frame, the spatial profile of an electron has a radiating electric field as an electric charge. This means that the electric charge of an electron is the time average of varying electric fields in the internal frame of the electron, as discussed in Chapter 1. While the flying electrons have velocity v_t, the time scale at the internal frame is longer than the value in the laboratory frame as $\Delta t' = \gamma \Delta t$, where $\gamma = [1-(v_t/c)^2]^{-1/2}$. As discussed previously, if the flying corpuscle electron as electromagnetic pulse at the laboratory frame and the Δt or $\Delta\omega$ is the time duration or frequency broadband of electron pulse, then the electron pulse front is the spatial profile (or contour) front of the corpuscle electrons and the ensemble of the electromagnetic wave fields of the dynamic vorticity electromagnetic field of photon would have the components of the dynamic electromagnetic fields of the pulse at the time duration $\Delta t = \Delta t'/\gamma$ of the pulse at the microscope frame. This may be the definition of an electron pulse. Therefore, for transmission electron microscope, the electron beam as current of the electron charges would be the sequential flow of electrons with spatial $\Delta x \Delta y \Delta z$ or time Δt separation, and z-axis is the flying direction, which should be the time axis. These parameters are determined by the emission current of the electron gun and illumination electromagnetic lenses. This ensemble of flying charge corpuscles is a beam as flying bees swarm with a spatial enveloped contour. The flying direction (as longitudinal) profiles in front of the electron beam can be seen as a beam front. If the velocity of each corpuscle electron is same or if $\Delta(v_t/c) \ll v_t/c$, then the beam front is a two-dimensional (2D) plane. The transversal profile of the beam can be seen as the shape of the beam. As mentioned before, at the microscope frame, the kinetic energy of a corpuscle electron would be $\varepsilon_{kin} = (\gamma-1)\, m_0c^2$ and the kinetic energy varies with γ. At an electron gun, the electric field having symmetric distribution and spatial electric field gradients that make the γ varies with the path of the electron as $\gamma_2 - \gamma_1 = eE/\varepsilon_0(r_2 - r_1)$. At the cross-sectional plane of electrons exiting from the gun, the velocities, v_t/c, of the escaping electron have the distribution with δt_i time delay that makes the beam wavy front and the beam composed of the flying corpuscle electron pulses can be described by the Liouville theorem at position-momentum phase space. For transmission electron microscope, the continuous flying corpuscle swarm can be visualized as the plane wave with longitudinal wavelength as $\lambda_{beam} = 2\pi/\kappa_{||}$ and the transversal profile is L or $\lambda_T = 2\pi L = 2\pi/\kappa_\perp$. The temporal stability of the gun's static electric field is higher than 10^{-10}, which means the γ variation may be ignored. As discussed before, each corpuscle electron has almost paralleling de Broglie's wave vector $\mathbf{k}_{||} = [\mathbf{k}^2 - (\mathbf{k}_\perp)^2]^{1/2}$, and the electron beam has plane wave with $\lambda_{||} = 2\pi/|\mathbf{k}_{||}|$, but $\lambda_{beam} = 2\pi/\kappa_{||}$, where $\kappa_{||} = 1/\delta z$ and $\delta z = v_z\delta t$. However, the λ_{beam} is the temporal average of the de Broglie wave vector $k_{||}$, $<2\pi/|k_{||}|>_{\delta t}$. Due to the spatial-temporal coupling of the de Broglie wave vector, the spatial contour in the transversal plane and the time delay distribution of the corpuscle electron in the longitudinal direction would dominate the beam front geometry that could be manipulated by the external force, for example, magnetic lens. It is clear that there are two factors to steer the beam front geometry. (1) The corpuscle electron as the flying charge particle can be manipulated by the electromagnetic external field existed at its loci of path of trajectory of the flying corpuscle causes the velocity deflection from the original direction, resulting in the $\gamma = [1 - (v_t/c)^2]^{1/2}$ factor modification. (2) Simultaneously, the beam front geometry would be manipulated as from the plane to curvature due to group velocity direction of the internal electromagnetic fields due to phase velocity variations, $\delta v_g = c^2/\delta v_{ph}$ and $\delta v_{ph} = \partial\varphi/\partial t$, which emanated from the interaction between the environment medium fields, where the corpuscle electron is traveling, and its internal vorticity electromagnetic fields. This causes the individual corpuscle velocity being tilted and/or curved, resulting in the beam front from the flat plane to the curve or corrugated. This would cause repartitioning of the original beam to separate into different directions as the diffraction beam.

The $\gamma = [1 - (v_t/c)^2]^{-1/2}$ factor modification entails on the temporal characteristics of the dynamic process of the electromagnetic energy flux at the corpuscle electron frame that results from the rotation frequencies of the vorticity field of the photon at the corpuscle electron frame are modified or usually the said phase rectifying the electron may be expressed as $\varphi = \pm (c/\ell) \int_{t_1}^{t_2} [1 - (v_t/c)^2]^{1/2} dt'$, where $\ell = \hbar/m_0c$ is the radius or wavelength of the Compton and $\tau = \ell/c = \hbar/m_0c^2$ is the temporal periodicity of rotation of the vorticity field of a photon at the corpuscle electron frame and the τ is called a "proper time" which is same in different inert frames. The time t′ is the transient time of circulation in the Compton rotation as rotation angular $\theta \to t'/\tau$. As discussed previously, the vorticity field moving paths of photon in the corpuscle electron frame may show as a helix-type line along the particle's ultimate border, which is the contour of the corpuscle frame, and the moving velocity at any point along the helix-type line would be the speed of light. For Minkowski's space, this helix-type motion will be the cylindrical helix, but the "world line" is depicted by a point of the particle's border instead of the particle itself. The line element of its path δs would be proportional to a differential of the action function as follows:

$$mc\delta s = -\hbar\delta\varphi = mc^2\delta t\left[1-(v_t/c)^2\right]^{1/2} = \delta S = \hbar(\partial\varphi/\partial t)\delta t + \hbar(\partial\varphi/\partial z)\delta z \rightarrow$$

$$\hbar\omega\delta t + \hbar k_n\delta x_n = -mc^2\delta t\left[1-(v_t/c)^2\right]^{1/2} \approx -\varepsilon\delta t + p_n\delta x_n$$

Due to $\varepsilon = \left(mc^2 + mv_t^2/2\right) = |\omega|\hbar$ and $p_n = \hbar k_n$, which is the de Broglie's relationship, that is, the free quantum particle with positive parity function expression as usually said, the wave function $\psi = \psi_0\exp[i(p_n x_n - \varepsilon t)] = \psi_0\exp[i(k_n x_n - \omega t)]$. It is clear that for immobility in space, the particle's 2D rotation cell must be permanently pumped over as a flicker, which may be a Möbius strip loop, the flickering frequency value $|\partial\varphi/\partial t| \equiv \omega_0$ determining the particle's energy at rest $\varepsilon_0 = m_0c^2 = \hbar\omega_0$. These discussions indicate that the electron at the laboratory frame is a charge and mass particle, but it has hidden universal spatial-temporal dynamics of rotation and circulation of electromagnetic vorticity fields of a photon.

The corpuscle electron beam may be expressed as a space and time separable function as $\Psi(x,y,z=v_t t) = \Psi(x,y)\Psi(t=z/v_t)$, but the corpuscle electron pulse is a spatial-temporal coupling function that intimidates the electron beam geometry's front shape to closely relating the group velocity orientation of individual corpuscle electron pulse and the beam front tilting and/or curved result splitting the beam front into scattering beams with different angles corresponding to the original propagation direction. Therefore, the transmission electron microscopy could be divided into two groups: (1) electron optics or quantum electron optics, which focus on how to steer the path of trajectory of the flying electron as a charged particle or how to manipulate the transversal shape and beam front geometry of the electron beam by magnetic lens for different illumination modes and the final magnification of the lens system. Recently, the interaction between the stronger laser beam field and the corpuscle electron pulse has caught the attention of researchers and the transversal shape of the electron beam is manipulated for detecting the transversal de Broglie's wave vector or momentum that coded the history of the interaction on its trajectory. (2) Using different illumination modes to steering the electron beam traversing through samples such as plane beam front paralleling entrance into the sample as Iijima and Cowley modes; convergent curvature beam front with different curvature angles traversing in the sample; and integrating the convergent curvature beam front with the objective lens defocus to observe the near-field image of each splitted component beam. As discussed previously, the group velocity of the corpuscle electron pulse such as $v_g = \sum_n |h_{n(k)}|^2(\omega/k_n)x_0; v_n = (\omega/k_n)x_0$ carries the accrued information of nanoscale action of the corresponding corpuscle electron with the sample, where $|h_{n(k)}|^2$ is the parasite at the intensity, and k_n is the evanescent vector of the corresponding corpuscle electron pulse. At the near field, the individual group velocity orientation changing the corpuscle electron would locally induce the tilted or curved front of the corpuscle electron pulse, and this parasite information will lose at far field at the back focal plane of the objective lens.

4.2 What you see at the screen or CCD device?

For a corpuscle electron pulse (which may be visualized as the torus pulse), the space-time is coupled with the relation, $(\omega_{||}+\omega_{\perp})(\omega_{||}-\omega_{\perp})=(ck)^2$. The electric field of this pulse is $E(x,y,z,t)=E_0\delta(\varphi)\delta(z-v_t t)=\sum_n E_0\exp\{i(k_n r-\omega_n t)\}\delta(z-v_t t)=q^-(x,y)$ $\delta(z-v_t t)$, where q^- is the electric charge of an electron. The electron current density is $I=Nq^-v_t$, for example, for transmission electron microscope at 200 kV, $v_t=0.695c$ (c is the speed of light). If the typical current is 1 nA, which is equivalent to 6.25×10^9 electrons per second, then $Nq^-=I/v_t$ being linear density of electrons would be 30 electrons/m (nA), the average distance between two consecutive electrons along the propagating direction is $<\Delta z\geq 3.3$ cm. This implies that electrons in transmission electron microscope for a current of 1 nA would consecutively impact on the screen of the observer as an electromagnetic pulse at a time interval $\Delta t_{min}=16$ ns. The electron density distribution at 2D plane would determine the brightness distribution at the screen of the observer. Therefore, the imaging pixels on the screen directly correspond to the density $\Delta Nq^-/(\delta r(x,y))^2\delta t$ of electron charge at a spatial point r(x,y). The contrast of imaging pixels depends on the ratio between the background spatial-temporal averaged electron density and the given spatial pixels as $\xi(x,y)=\tilde{I}_{max(x,y)}-\tilde{I}_{min(x,y)}/(\tilde{I}_{max}+\tilde{I}_{min})\approx<\Delta Nq^-(x\pm\delta x,y\pm\delta y)>_t/<N$ $q^-(x,y)>_t$. As discussed previously, the two-time correlation function of corpuscle electron pulses allows one to determine whether the electromagnetic energy pulses in the ensemble as beam are coherent. The charged corpuscle electron with spatial contour scale of Compton wavelength $\ell=h/m_0c$ is anti-bunch, which means it is impossible to detect two charged corpuscle electrons at exactly the same time and location. As all measurements take some finite time, it is only possible to show that the coherence is deduced for small time duration τ. For the temporal correlation function of an ensemble of corpuscle electron pulse beam

$$\mu(\tau)=\left(\iint|<E(t_1)E^*(t_2)>|^2dt_1dt_2\right)\Big/\left(\int<E(t)E^*(t)>dt\right)^2\quad\text{where } t_2=t_1+\tau,$$

The width of the temporal correlation function $\mu(\tau)$ is the reciprocal of the width of the spectral distribution, which implies the spectral dispersion decrease coherence of the beam. We have previously mentioned multicoherence function as $\gamma(r_1,r_2,t_1,t_1)=$ $\Gamma(r_1,r_2,t_1,t_2)/[I_1(r_1,t_1)I_2(r_2,t_2)]^{1/2}$ with $\Gamma(r_1,r_2,t_1,t_2)=(1/N)\sum_{n=1}^{N}(E^*{}_n(r_1,t_1)E_n(r_2,t_2)$

Then the contrast of the image of electron microscopy at any spatial point is the ratio of the temporal integrated electron numbers at this point with observed spatial average value. Therefore, increasing the temporal interval τ of consecutive corpuscle electron pluses, which means decrease the frequency different, $\Delta\omega_i$, between the different flying corpuscle electron pulses, that means properly decreasing the electron beam current value would increase the contrast of image of electron microscopy.

4.2.1 Electron beam's longitudinal coherence

As mentioned before, the corpuscle electron beam transmitting direction in the transmission electron microscope is the optical axis or z-axis, on which is also the time axis, $t = z/v_t$. This direction is also the propagation direction of the de Broglie wave of the corpuscle electron, which is the longitudinal wave vector direction, $k_{||}$, if it is parallel to the beam front transmitted direction. The coherence along this direction is the temporal coherence. The coherence time is $t = 1/\Delta\omega$ that indicates if every individual corpuscle electron pulse has same spectral frequency $\Delta\omega = \omega - \omega_0$, the beam has complete coherence, which means the beam coherence is same as the individual corpuscle electron pulse. If the spectral frequency is different, $\Delta\omega = \omega - \omega_0$ increases, then the coherent time would decrease. Electron beam traversing over some material may result in increasing longitudinal coherence, inducing the beam intensity to be annihilation as usually called "extinction."

4.2.2 Transversal coherence

If the individual corpuscle electrons have same energy $\mathcal{E}_n = h\omega$ and flying parallel to each other, then transversal coherence can be used to measure the propagating corpuscle electron pulse wave vector spread $\Delta k_{||}$. So, we could define the transversal coherence length $\ell_c = 2\pi/\Delta k_{||} = 1/\Delta\theta$, which indicates that the transversal coherence of the ensemble of corpuscle electron pulses is dominated by the angular divergence of wave vector $k_{||}$ of the ensemble corpuscle electrons. That is why defocus of illumination magnetic lens with longest focus length obtains the plane beam front with longest transversal coherence length, ℓ_c, for high-resolution image of electron microscopy. Increasing the convergent illumination angler would decrease the transversal coherence, ℓ_c, which results in the corpuscle electron pulses that carry more information of the electromagnetic field of the sample.

4.3 Symmetric spatial-temporal periodicity of the electromagnetic field of sample detected by the ensemble corpuscle electron pulse beams as Talbot's effect

The enigmatic nonstoichiometry of oxides was deciphered by exposing the symmetric spatial periodicity of atom's groups of inorganic compounds, using high-resolution electron microscopy (HREM) during later decades of the last century, but how to understand the obtained image contrast is still debated.

Talbot reported his observation of axial self-imaging of a grating in an optical beam field endowed with a periodic transversal spatial profile of the grating. Such a periodic field of optical beam first undergoes diffractive spreading in space of the grating. The resulting initial periodic optic field structure of the beam is progressively lost, but the original periodic profile of the beam is modified to be the grating spatial periodicity field and to be revived at planes separated by the Talbot distance $z_{T,x}$. This is called as the spatial Talbot effect, in which the spatial periodic structural features of grating exhibit on the transversal intensity periodicity of the optic beam field revived at the plane of propagating distance $z_{T,x}$. Similarly the temporal Talbot effect exists, where dispersive spreading in time replaces diffractive spreading in space. Pulses in a periodic pulse train first spread and overlap temporally, followed by subsequent revivals of the original structure in the time domain at planes separated by the temporal Talbot distance $z_{T,t}$.

The corpuscle electron as an energetic particle with mass, m, whose initial kinetic state may be given by a one-dimensional periodic wave function as follows:

$$\Psi\,(t=0, x+d) = \Psi\,(t=0, x)$$

where d is the period of the wave function of the corpuscle electron. As the wave function of the corpuscle electron is spatially periodic, then it can be spatially decomposed in a Fourier series:

$$\Psi\,(t=0, x) = \sum_{j=-\infty}^{\infty} \Psi_j \exp\,[2\pi\, ijx/d]$$

If the Hamiltonian of a free corpuscle electron is $H = -\hbar^2\partial_x{}^2/2\,m$, the time evolution of this wave function Ψ is

$$\Psi\,(t, x) = \sum_{j=-\infty}^{\infty} \Psi_j \exp\,[2\pi\, ijx/d] \exp[-i\pi\, j^2 ht/md^2]$$

which is periodic in time with a period, $T_t = md^2/h$, and accompanying a spatial shift. The time period may be referred to as the Talbot time $T_t = md^2/h$, where d is the spatial periodicity of the corpuscle electron.

At the Talbot time, wave function of the corpuscle electron is

$$\Psi(T_t, x) = \sum_{j=-\infty}^{\infty} \Psi_j \exp[2\pi i j x/d] \exp[-i\pi j^2] = \sum_{j=-\infty}^{\infty} \Psi_j \exp[2\pi i j x/d] \exp[-i\pi j]$$
$$= \sum_{j=-\infty}^{\infty} \Psi_j \exp[2\pi i j(x - d/2)/d] = \Psi(0, x - d/2)$$

This result means that the kinetic state of the corpuscle electron with internal periodic motion would have the state with half of its intrinsic spatial period at the Talbot time. The spatial-temporal periodicity is the physical origin of the Talbot effect.

If we understand the ensemble of corpuscle electron pulses beam with energy $\mathcal{E} = h\omega$ as a field having spatial-temporal periodic structure, then we describe this beam as $\Psi(x, y, z = v_t t) = \exp[i(kz - \omega t)]\Psi_0(x, y, z)$ with transversal profile L and spatial periodicity $\kappa = 2\pi/L$ in free space. The $\Psi_0(x, y, z)$ is the envelop function describing the transversal profile with L and longitudinal scale as time axis with $t = z/v_t$. At the laboratory frame, spatial profile scale of each individual corpuscle electron can be seen as the Compton length $\ell_{comp} = \hbar/\gamma m_0 c$ that is related to (v_t/c), at the transversal plane (x,y). But the time axis, t, at the direction of z-axis is related to the velocity, v_t. For the ensemble of corpuscle electron beam, the beam front shape $\text{F}(x,y,t_i)$ is dominated by the velocity distribution as $\text{F}(x,y,z/v_t)$. If we consider the corpuscle electron beam as the pulse train, then the pulse front simultaneously arrived at the corpuscle electron trace surface and the duration of the pulse is the time interval between the two consecutive trace surfaces. The frequency interval of the front and back of the pulse is $\Delta\omega = 1/\Delta t_n$, and the time coherence is $\Delta t_n = 1/\Delta\omega$. The transversal coherence is $\ell_c = 2\pi/\Delta k_{||}[1 + (\kappa - k)]$. If the beam front vector is $\kappa = k$, then the transversal coherence of the beam is same as the individual corpuscle electron pulse, $\ell_c = 2\pi/\Delta k_{||}$. The transversal profile dimension of the beam would be $N\ell_{comp} = L$.

Crystalline sample may be seen as the stationary spatial periodic atom's grating in which the nucleus of the atoms built the wall of the grating and the circulating electrons around the atom's nucleus form the electromagnetic fields in the chink of grating because the nuclear scale of an atom is about 10^{-15} m or $\approx 10^{-3}\ \ell_{comp}$, then the chink space of atom's grating having atomic scale, $\sim 10^{-10}$ m, is generally a space with the electromagnetic field. The flying corpuscle electron having 10^{-13} m transversal scale can transmit it easily and the longitudinal electric field in the chink of the grating may modulate the flying velocity of the corpuscle electron, which induces the kinetic energy variation of the corpuscle electron that may result in the Pendellösung effect, but the transversal field may rectify the velocity direction of the corpuscle electron as momentum exchange.

As mentioned previously, the Talbot effect has been observed in optics and interpreted in various domains of wave physics, including quantum mechanics and electromagnetism, in general. The original Talbot effect was first observed and explained in the spatial domain. This was later observed in the temporal domain and in the dual Fourier domains of spatial and temporal frequencies as angular and spectral Talbot effects, respectively. The most appealing features of the Talbot effect involve the possibility of creating an exact replica of a given periodic pattern (self-imaging, or integer Talbot effect) or the division of the periodicity of the original pattern by an integer factor (fractional Talbot effect). These properties may be used to understand the corpuscle electron's electromagnetic pulse train with high repetition rate as the Talbot effect in the temporal domain. Essentially, the Talbot effect originated from periodicity of fields in which the field evolution in space and time is correlated by virtue of the field structure itself. The key to realizing this space-time Talbot effect is the identification of a novel pulsed optical beam structure having spatial and temporal degrees of freedom to be inextricably linked. The intrinsic electromagnetic field of the kinetic and dynamic motions of the vorticity field of a photon in a corpuscle electron frame has inextricable linking between the circulating frequency ω and the rotation radius, r (x,y), by $r(x, y, t)\omega(\tau) = c$ in the corpuscle electron frame. For microscope frame, there is a relation $\hbar\omega = [(\hbar\omega_0)^2 + (\hbar ck)^2]^{1/2}$, which shows the inextricable link of spatial and temporal degrees of freedom, as shown in Figure 3.2. This makes the periodic field of a corpuscle electron pulse to be able to revive at z_T of the propagating axis. This revive includes the transversal spatial profile, ℓ_{comp}, and longitudinal scale $z_{T;t}$. This makes people to propose the electron as a wave packet pulse. The ensemble of flying corpuscle electron as beam impacts to the crystalline sample, which is stationary atom's grating with slits or chinks. The flying corpuscle electrons as quantum particles travel through gap space of the slits or chinks. The traveling time in the gap space of the slits or chinks will dominate the interaction between the flying corpuscle electron and the electromagnetic field in the gap space of the slits or chinks. As mentioned previously, the Talbot time, $T_t = md^2/h$, indicates the kinetic state shift of the half-period of space of a quantum particle. We use the kinetic energy $\varepsilon_k = p^2/2m = h^2k^2/2m$ to induce the momentum variation as a frequency for recoil momentum as $\omega_r = hk^2/2m$ and the depth of slit or chink (also as lattice depth) as $q(t = z/v_{in}) = V(t)/4h\omega_r$ in which the V(t) is the potential in space of a slit or chink, and the $q(t = z/v_{in})$ may express variation of the ratio of potential and recoil energy during transient time. The time $t = z/v_{in}$ is the traveling time in the gap space of a slit or chink with its incoming velocity, v_{in}. The duration of traveling in a slit or chink would be $\Delta t = z_{out}/v_{out} - z_{in}/v_{in}$. The spectral phase changes $\Delta\varphi = (\partial\varphi/\partial t)\Delta t$ during traveling. For transmission electron microscopy, the electron beam's transmitting time of sample is $t' = \Delta z(x, y)/v_t$, where Δz (x,y) is the thickness of the sample. For 200 kV, electrons transmitting through 0.1 and 10 nm thick crystals need about 6×10^{-21}s and 6×10^{-19}s, respectively, which indicates that while two order thicknesses increased, the transmitted time duration also gain two orders. The Talbot time $T_t = md^2/h$ may be understood as the time duration required for rotating half-spatial periodicity around the center of mass of electromagnetic energy flux of vorticity field of a photon in the corpuscle electron frame, which observed at the laboratory frame, due to $mc^2 = h\omega \to h/t = m(s/t)^2 \to ms^2 = ht \to T_t = ms^2/h$, if s is the rotation circumference as Compton length ℓ_{comp} as spatial periodicity, d, of the circulating photon at corpuscle electron frame or wave function $\psi(t = 0, x) = \psi(t = 0, x + d)$.

This implies that the Talbot time, $T_t = md^2/h$, in the laboratory frame is related to the corpuscle electron internal clock's timescale, $\tau = d/c = h/mc^2$, of the corpuscle electron, and d is related to the Compton wavelength, $\ell_{comp} = h/mc$, of a corpuscle electron. As the basic physical law, the superposition between the oscillators has to manipulate in a periodicity of spatial-temporal domain. Therefore, the Talbot time T_t and the Compton wavelength ℓ_{comp} are the basic physical parameters for flying corpuscle electron in vacuum and medium. In the laboratory frame, a propagating distance, δz, of a corpuscle electron may be expressed as $\delta z = v_t T_t \rightarrow \delta z = v_t md^2/h = v_t/\omega_t$, which means the displacement, δz, of a corpuscle electron in the propagation direction is the ratio of transient linear velocity, vt, and transient frequency of the rotated electromagnetic energy flux.

While the propagating direction of the ensemble corpuscle electron pulse train is collimated with the axis of high-symmetric low-index zone of a crystalline sample, the grating formed by the symmetric atom planes would have slits or chink tunnels with $q(t = \delta z/v_t) = V(t)/4\hbar\omega_r$. While a particular individual corpuscle electron with momentum $p = hk_{||}$, identified with its electromagnetic field, will travel through these slits or chinks and during its traveling time in a slit or chink, the corpuscle electron's electromagnetic field will interact with the field in the slit or chink constructed by the atom planes, resulting in the momentum exchange between the traveling corpuscle electron and the atom planes of the crystalline sample. The traveling temporal duration in a slit or chink would dominate the momentum exchanged amount, and this determines the lateral momentum exchange, or, equivalently, in a lateral spatial shift of the traveling beam. As well known, the Bragg law is $2d\sin\theta = n\lambda$, and now wavelength is the de Broglie wavelength λ_{dB}, θ is the angle between the traveling quantum particle momentum and the atoms' lattice planes of the grating. The quantum particle $2d\sin\theta = n\lambda_{dB}$ is equal to $p\sin\theta = n\hbar k$ implies that the components of the corpuscle electron's momentum along the lattice axis are integer multiples of the corpuscle electron momentum. If we accept these ideas, we may define that the "grating momentum" $\hbar G$ is equal to $\hbar k_L$ or $d = \lambda_L$, and the Bragg diffraction is a quantized momentum exchange between the incoming corpuscle electron pulses and the scattering crystal, which can occur only in multiples of the "grating momentum" $\hbar \mathbf{G}(\mathbf{G} = 2\pi/d)$. In general, the potential being periodic modulation in the time domain with frequency ω_m, under certain condition, can exchange energy quanta in multiples of $\hbar\omega_m$ with the traveling corpuscle in the slit or chink of grating. For standard scattering process, the time duration of the corpuscle electron pulse in a slit or chink of grating would make significant difference in the scattered direction. If the traveling time duration is smaller than the Talbot time, for example, $0.5T_t$, the scattered angle is very small and can make every orientation matching with the incoming transversal de Broglie wave vectors. While the traveling time duration is larger than the Talbot time, for example, $0.5T_t < T_t < 5T_t$, the crystal is thick enough to define a sharply fixed orientation of the grating vector or the potential endowing the recoiled frequency ω_r to match with the de Broglie transversal wave vector $\hbar k_\perp = \hbar k_{||}\sin\theta$, which selects the incoming pulse of the ensemble corpuscle electrons having the inclined angle $\sin\theta = k_\perp/k_{||}$ to be scattered. In other words, the beam front of the ensemble of the flying corpuscle electron pulses would be modified by the field of the slit or chink. A particular individual's internal frequency of the corpuscle electron would vary with its loci of the traveling path causing the chirps of the de Broglie waves in the tilt or curve of the individual corpuscle electron corresponding to the local beam front being tilt or curve, thereby forming the Talbot spatial-temporal effect. Because the amount of tilt and curve of a local corpuscle electron pulse determined by the ratio of transient velocity and recoil frequency depends on the local atom characteristics of the crystalline sample. Therefore, if using the Talbot time $T_t = md^2/h$ to scale the thickness, we may have three regions: (a) $t \le 0.5\ T_t$ is usually called Raman-Nath region, which is very thin (for TEM it is about a few angstroms); (b) $0.5T_t \le t \le 5\ T_t$ is called the Bragg region; and (c) $t \ge 5\ T_t$ is called the Pendellösung region or thicker region. As T_t contains mass of the electron that is related to the energy of an electron or acceleration voltage, then the Talbot time is related to the acceleration voltage of the electron gun. The Talbot time actually indicates the corpuscle electron's spatial-temporal characters as an ensemble of rotating the vorticity field of photons in the laboratory frame due to spatial-time coupling of the internal dynamic motion of the corpuscle electron.

For high-resolution transmission electron microscopy (usually said as HRTEM), it has to take the lowest index zone of a crystalline sample to be collimated with the direction of propagating velocity of the ensemble of corpuscle electron pulse beam. In this condition, the slit or chink (or tunnels) of grating constructed by atom planes aligns with the momentum direction of the corpuscle electron $\mathbf{p_0} = \hbar \mathbf{k}_{||}$, and while the individual corpuscle electron travels a Talbot time, T_t, the corpuscle electron pulse would transversally shift $d = \ell_{com}$, which is the space periodicity of a corpuscle electron, resulting in the angular deflection $\delta\theta$. This angular deflection $\delta\theta$ causes the momentum variation $\Delta p = \hbar k_{||}\delta\theta = \hbar k_{\perp} \approx \Delta\omega_r/c$. As time is gaining in the gap space of the slit or chink of the grating, the corpuscle electron travels deeper and more shift of transversal direction in a slit or chink of the grating and if the momentum transferred is as $\hbar k_{in} - \hbar k'_{out} = \delta p$, which may be described as the recoil frequency, $\delta p = \delta\omega_r/c$. If the potential V(t) in the tunnel of grating of atom planes is expressed as $V(t) = \sum_{\Delta z} (h\omega_r \Delta z_i / v_t)$, then the basic characteristics of Bragg diffraction is a quantized momentum exchange between the incoming corpuscle electron and the scattering lattice of the crystal, which can occur only in multiples of the grating momentum, $\hbar G$. The grating momentum G is the grating vector of the crystal having absolute value $|G| = 2\pi/d_{hkl}$, which is inversely proportional to the grating periodicity d_{hkl}.

For HRTEM, the depths of the tunnel of grating constructed by atom planes are closely related to the thickness of the crystalline sample if the transmitted time, t, is less than 0.5 Talbot time. It is called as the Roman-North zone or very thin region as a few atoms. In this region, the incoming flying corpuscle electron can be continuously scattered to every orientation that results in the scattered evanescent wave interference of an atom. If the transmitted time, t, is $0.5T_t \le t \le 5\ T_t$, or called the Bragg region, the outgoing corpuscle electron, $p' = \hbar k'$, only can have $k' = k_{||} \pm G$. In the Bragg region, the grating is spatially width enough to define a sharp fixed orientation of the grating vector and has selected recoiled frequency ω_G to make the energy flux flowing along the boundary of the first Brillouin zone of the crystal. Due to the fixed momentum exchange between the incoming corpuscle electron pulse train and the fixed discrete momentum, $\hbar G$, of crystal lattice, the periodicity of spatial crystal would encode on the outgoing ensemble of the corpuscle electron pulse train to form the self-image of the grating, which endows the corresponding spatial periodic assemblies of the ensemble of corpuscle electron pulse beam with tilting and curving. It is clear that the HRTEM image is the self-image of spatial periodicity of the grating, which exhibits on the transversal density distribution of the electron beam at the spatial Talbot distance $Z_T = d^2/\lambda$, but not the projection of the atom's column of the grating. To analyze the fingerprint of interaction between the traveling corpuscle electron and the slit or chink of the grating built by atom planes is the way to decipher the spatial periodic structure and atom planes physical parameters of the crystalline sample. As discussed previously, the depth and wideness of a slit or chink of grating built by atom planes may express the time-spatial domain for the interaction between the traveling corpuscle electron pulse and the atoms of the sample. The Talbot time, $T_t = md^2/h \approx v_t/\omega_t$, which contains the velocity ($m = \gamma m_0$) and the transversal dimension is scaled as the Compton wavelength, $\ell_{comp} = h/mc$. Figure 4.1 shows the field of single corpuscle electron, in which the vorticity field of the magnetic part of the rotation electromagnetic field of a photon is perpendicular to the electric field of corpuscle electron and the Talbot time may be seen as the temporal periodicity of rotation, and the transversal circulation circumference of the vorticity field is the spatial periodicity, $d = \ell_{comp}$. Δz is the displacement in a temporal period of the corpuscle along the propagating direction.

Then at a Talbot time, the circulating photon would shift a transversal spatial periodicity d and longitudinal displacement Δz. If we accept this picture of a corpuscle electron, then the spatial intervals between the corpuscle electron pulses should be larger than d and the overlapped temporal period between the corpuscle electron pulses should be equal to the Talbot time $T = md^2/h$, which dominates the temporal interval δt between consecutive corpuscle electrons. Therefore, we may understand the basic parameter of the ensemble of consecutive flying corpuscle electron pulse train or beam as τ and ξ. If the spatio-temporal intervals of the two corpuscle electrons are in the range of the Talbot time and spatial period d, their interaction processes are the vorticity fields of electromagnetic action as Maxwell's laws, but if their spatio-temporal intervals are larger than the Talbot time and spatial period d, then their actions would be as charge particles with q_e/m, which is a physical entity of confined kinetic and dynamic electromagnetic fields of the circulating photon. At the first situation, we have to use the Maxwell and quantum electrodynamic theories to explain the observed experimental results, but at the second situation we may use the Maxwell Lorenz, Schrödinger, and Newton theories to understand the result. For transmission electron microscopy,

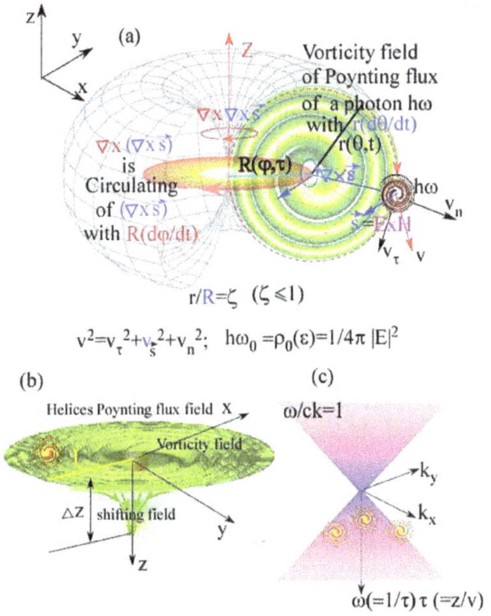

Figure 4.1: Corpuscle electron pulse model. (a) The vorticity field of Poynting flux of a photon with circulation, dθ/dt, rotated at plane R(φ,τ), that is, the energy flow as $\nabla x(\nabla x\mathbf{S})\ ||\ \mathbf{v_t}$ with linear electric field along the propagating direction (z-axis). (b) Corpuscle electron pulse has vorticity magnetic field as spin, and linear electric field and vorticity electric field as $1/c(\partial B/\partial t) + \nabla x E = 0$. The energy of a corpuscle electron is the vorticity field energy, $h\omega_0 = 1/4\pi\,|E|^2$ and kinetic energy of the corpuscle electron concentrated around the linear electric field. (c) The kinetic path of the circulating photon appears on the surface of the optic cone. The photon at each track of the periodic path of kinetic circling of a photon has the speed of light.

the second situation may usually be used as the group velocity of a corpuscle electron pulse plays the main rule for elementary tilt and curve of front of the ensemble corpuscle electron pulse beam. In Figure 4.2, we use these ideas to understand the HRTEM image formation. When the thickness Δz is very thin (Raman-Nath region) as $\Delta z < 0.5\ T_t v_t$, the corpuscle electron travels through atoms, which means the corpuscle electron overlapped with the atoms spatially, and the interaction between the corpuscle electron and the atom is the action of the internal electromagnetic field of a corpuscle electron with the field of atom, as shown in Figure 3.4. This may be the observed single-atom image at the edge of the crystal on the screen. At the Bragg region ($0.5T_t \leq t \leq 5\ T_t$), the width of the silt or chink of the grating built by atom planes in the lowest index zone of a crystal, which has the highest symmetry width space, is large enough compared with the Compton wavelength $\ell_{comp} = h/mc$. The ensemble of the consecutive flying corpuscle electron pulse beam passed through the grating is reconfigured by grating to create the new ensemble spatio-temporal configuration having the spatial periodicity of the grating. This is revived in the space plane at the Talbot distance $z_T = nd^2/\lambda_{dB}$, along the propagating direction (time axis) or z-axis as the Talbot self-image of the grating, which is shown in Figure 4.2. The contributions of the individual slits of the grating to the intensity in the near field have quadratic phase dependence in the paraxial approximation, while in the time domain, the Talbot effect arises typically in dispersive lines that imprint a quadratic spectral phase on the propagating pulses: the curvature of the parabola of the phase sets directly the specific integer or fractional Talbot pattern that is observed at the system output.

Since the Talbot image is the self-image of the grating, which is after a free propagation over the longitudinal distance ℓ within the slit opening of the grating and has imprinted a complex phase to account for the interaction potential between the grating wall and the beam corpuscles. Then the beam corpuscles fly into the vacuum and propagate at the free space in the near field beyond grating. In the case of the translation periodic grating with period d, the density distribution of the assembly of flying corpuscle electron pulses takes the form of the grating transmission profile wherever the distance ℓ is an integer multiple of the Talbot length $\ell = n(d^2/\lambda_{dB})$. The density distribution of the assembly of flying corpuscle electron pulses having angular divergence with the period d of the grating would propagate from the Talbot plane into the far field that is identical to the intensity diffracted in the far field. The intensity diffracted in the far field is simply as

$$I(k_x) \propto \sum\nolimits_l \delta((k_x - lK_0) = \sum\nolimits_n \exp(i2pnk_x/K_0),$$

If we use the Wiener-Khinchin theorem for the intensity at $z = (p/q)z_T$, then we can write:

$$I(k_x) \propto FT[E(x, (p/q)z_T)^* E^*(x, (p/q)z_T -)] \propto \sum_n \exp(i2pnk_x/qK_0) \sum_l r_l r^*_{l+n}$$

The autocorrelation $\sum_l r_l r^*_{l+n}$ vanishes except when n is a multiple of q. The sequence of the phases occurring in the Talbot images is an example of delta-correlated discrete sequences that can form the interference at the near field.

The classical Talbot effect is a near-field effect. Indeed, any interferometric method requires mixing the Talbot image with a reference light beam, for example, the incident light field. The HREM image is mostly created by the transmission beam and diffracted beams, for which they fusion each other at the Talbot image intensity behind the sample and separated at the back focal plane of the objective lens. Using convergent beam diffraction, Cowley mixes the transmission and diffraction beam disks to obtain the high-resolution image of $Ti_2Nb_{10}O_{29}$ along [100] of crystal in HB5 scanning transmission electron microscope.

However, the Talbot image is located near the exiting surface of the sample, and for capturing these images, Iijima had adjusted the defocus of the objective lens for steering the objective plane to the spatial Talbot distance z_T, which is called the series

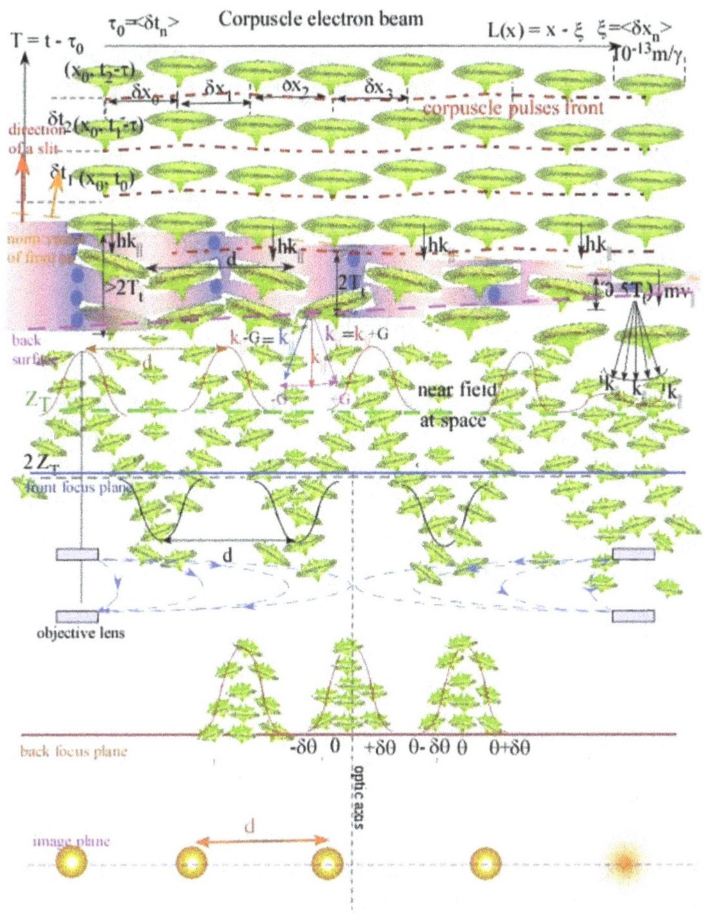

Figure 4.2: The ensemble of flying corpuscle electron pulse beam transverse a grating build by atom planes to create the spatial and temporal Talbot self-image of the grating. The consecutive time interval of the flying corpuscle electron is δt, and the averaged consecutive time interval is $\tau_0 = <\delta t_n>$. The transversal spatial interval is δx_n and the average transversal spatial interval is $\xi = <\delta x_n>$. Every corpuscle momentum is $hk_{\parallel}$, and G is the momentum of crystal lattice, $G = 2\pi/d$ (d is spatial periodicity of the grating built by atom planes). T_t is the Talbot time, $T_t = md^2/h$ (m is mass of an electron, d is the spatial periodicity of a corpuscle electron at its internal frame, and h is Planck's constant).

defocus method or Iijima-Cowley method for HREM. The series defocus is necessary to get the correct Talbot image of the grating constructed by the atoms.

Intensity distribution of the Talbot image is dominated by the length of the slit ℓ_s, which gives the flying time duration $\Delta t_s = m_e \ell_s / p_z$ (where m_e is the mass of an electron and p_z is the momentum of the flying corpuscle electron in the propagation direction), and the potential in the slit, which is related to the spatial window, $\Theta = (d - b_w)/d$, of the silt (d is the period of the grating, b_w is the wall thickness of a slit of the grating, and Θ is the opening spatial rate of a slit of the grating). As mentioned before, the transmission function t(x,y) with $|t(x,y)| \leq 1$ describes the multiplicative modification of the incoming corpuscle electron pulse beam in the slit opening of the grating. In the laboratory frame, the corpuscle electron pulse is a charged corpuscle with electric field and spin, and the trajectory of the corpuscle electron may use the eikonal approximation to describe its kinetic motion and interaction with the potential. The grating and the combination of an absorption mask and a phase modification could express the grating transmission function as follows:

$$T(x,y) = |t(x,y)| \exp\left\{ -(im/\hbar p_z) \int dzV(x,y,z) \right\}$$

For a free-interaction grating slit without phase, $|t(x,y)|$ is equal to t(x,y), and for pure phase grating, $|t(x,y)|$ is one. While t(x,y) is non-zero within the slit opening of the grating, it imprints a complex phase to account for the interaction between the potential in an opening slit and the flying corpuscle electron. After a free propagation over the longitudinal distance ℓ_s, the transversal density of the corpuscle electron pulse beam is given by the marginal distribution

$$W(x,y) = \sum \sum B_m(m(L/L_T))B_n(n(L/L_T))\exp\{2pimx/dx\}\exp\{2piny/dy\}$$

where $B_m(\xi) = \sum b_j b^*_{j-m} \exp\{ip\xi(m-2j)\}$.

The character length scale $\ell_T = d^2/\lambda$ is called the Talbot length, and the transversal density distribution takes the profile of the transmission function form of the grating whenever the free-flying distance within an opening slit of the grating is equal to an integer multiple of the Talbot length, and the transmission transversal density has the periodicity of the grating as $\rho = |t(x + \varsigma(d/2))|^2 = |t(x,y)\exp\{-(im/\hbar p_z) \int dzV(x,y,z)\}|^2$. The transversal density period d/2 would emanate from the flying temporal duration with a Talbot time $T_t = md^2/h \approx v_t/\omega_t$. Within the grating passage time $t = m\ell_{slit}/p_z \approx m\ell_{slit}/hk_{||} \approx m\ell_{slit}\lambda_{dB}/h \approx m(\varsigma d)^2/h \approx \varsigma^2 T_t$, which is the multiple Talbot time. This is the transversal shift ςd that indicates the spatial dispersion, which means the angular dispersion of the incident flying corpuscle electron beam.

The spectral phase of the beam may express $\exp\{-i(k_{x,y})^2/k_z\}$ or $\exp\{-i(k_{x,y})^2/k_{||}\}$. Due to the grating of the crystal, the velocity of the corpuscle electron pulses is the group velocity of the corpuscle electron the grating passage time would relate to $1/\tilde{\upsilon} = dk_z/d\Omega$ and $k_2 = dk^2_z/d\Omega^2$ (group velocity delay parameter), and the temporal Talbot effect occurs when the field profile at z = 0 is periodic in time with period T. The temporal spectrum of this corpuscle electron pulse train is discretized at $\omega = m2\pi/T$ and the phase term $\exp\{i(1/2)k_2\Omega^2 z\}$ is responsible for dispersive spreading, which takes the form $\exp\{i2\pi m^2 z/z_{T,t}\}$. The corpuscle electron pulse train first disperses and the corpuscle electron pulses overlap temporally before it reconstitutes itself axially at planes separated by the temporal Talbot distance $z_{T,t} = T^2/\pi|k_2|$, and then it will have $\psi(\ell z_{T,t}, t) = \psi[0, t - \ell(z_{T,t}/c)]$. The temporal Talbot distance can be the difference to the spatial one, while the field profile of the assembly of the corpuscle electron pulses has the relation, $L = \alpha cT$, between the spatial period L and the temporal period T; then the spatial and temporal Talbot distance would be $z_{ST} = 2L^2/\lambda = T^2/\pi\,|k_2|$.

The spatial and temporal Talbot effect is based on the dispersion and diffraction proceeds in lockstep without external adjustments. Diffractive spatial spreading and dispersive temporal spreading are intrinsically synchronized, and a single Talbot length z_{ST} emerges. However, for HREM image, the contrast of the intensity of corpuscle electron ensembles is the time-average intensity, $I(x, y, z) = \int dtI(x, y, t)$, where $I(x,y,t) = |\psi(x,y,t)|^2$, because recording the intensity by a detector lacking temporal resolution is

$$I(x, y) = \sum |\tilde{E}_m|^2 + \sum |\tilde{E}_m\tilde{E}_{-m}|^2 \cos[4\pi m(x/L) + \varphi_m - \varphi_{-m}]$$

where $\tilde{E}_m = |\tilde{E}_m|\exp(i\varphi_m)$ is the discretized spectral field amplitude at $\kappa = m2\pi/L$. If we may interpret the L as the period of the grating, x is the transversal shift in the opening slit in the laboratory frame, and the phase κx is the passing of a slit-induced phase, and the temporal overlap of corpuscle electron pulses induces the phase difference $\varphi_m - \varphi_{-m}$. Therefore, the ensemble of corpuscle electron pulse beam that transmitted a periodic grating would be partitioned into two groups: one is spatial angular dispersion, which is $\sin\theta = n(\kappa/k_{||})$, and the other is temporal dispersion-induced transversal phase difference. While these two factors hold that the spatial-temporal coupling of the ensemble of the corpuscle electron pulse beam cannot be separated, the ensemble beam wave front and corpuscle electron pulse front are both curved by the same amount. But that is closely related to the opening window of a slit of the grating and the passage time, which means the large opening window of the high-symmetry Brillouin zones may see the white or dark dot image, and lower symmetry zone and/or thicker sample of crystal would not see any dot image but only interference fringes image for HREM. From this point of view, the so-called atom image may not make sense but the image is the "quantum carpet" of the crystalline sample.

Figure 4.3 shows the HREM image of TbO_2 image at [110] orientation. The white dots exhibit the opening slits of the fluorite structure along the [110] direction, and the opening slit or opening spatial tunnel hold by two metal atoms at horizontal and to two oxygens on vertical, that is large gap in the structure. while an oxygen atom mission this opening slit will get larger with modifying the spatial symmetry and change the potential in the slit results the additional contrast of the white dot, that may show the oxygen vacancies ordering in the TbO_{2-x} (Tb_2O_3) with fluorite structure.

Figure 4.4 shows that HREM image of same oxide TbO_{2-x} with different oxygen vacancies ordering ($Tb_{40}O_{72}$) at [112] orientation has low symmetry. The image demonstrates the (111) and (11–2) lattice fringes but it is not clearly showing the opening slits of the oxygen vacancies, but shows some effect of the interference.

It is worth to indicate that the depth (or Δz) and width (or window of a slit) dominate the Talbot images. The depth is related to the transmitting time and the width is

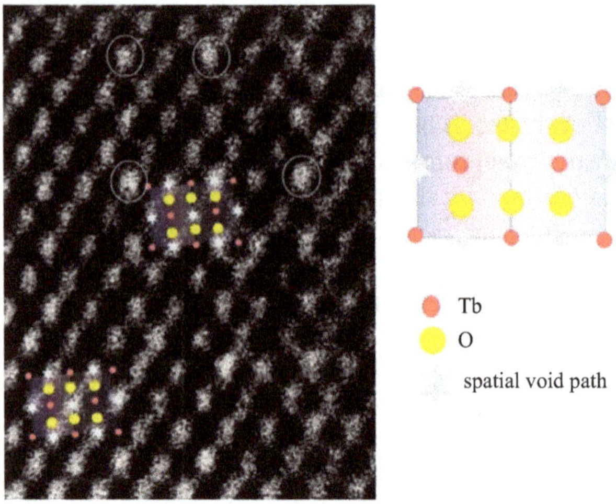

Figure 4.3: HREM image of TbO_{2-x} shows that the white dots are the opening slits of the grating constructed by Tb and O atoms in the [110] direction. The opening slits of the grating show white gray stars in the fluorite structure in the [110] projection. The gray contrast levels changed due to the neighboring oxygen mission that shows regularity due to its ordering.

related to the potential density distribution, which dominates the tilt and/or curve of the ensemble corpuscle electron beam front that influences the pattern of the Talbot image. These parameters induce the different features of the pattern of the Talbot image that is why the image simulation is necessary for the high-resolution image of transmission electron microscopy.

For the ensemble flying corpuscle electron pulse beam in transmission electron microscope, the sample, which does not matter amorphous, crystalline, or biomolecule, manipulates the orientation of the individual corpuscle electron pulse, based on the action function δS of its trajectory path, causing the beam front tilting and curving and partitioning the original ensemble flying corpuscle electron pulses into sub-ensemble corpuscle electron pulse train with angular dispersion as diffraction beams. The phase delay, which comes from the path's different induced time variations, results in the near-field interference as Talbot's self-image of the spatial stationary periodicity of the sample, which is the Fresnel effect. These Talbot self-images of the spatial stationary periodicity of the sample would be shown at several planes, $z_T = nd^2/\lambda_{dB}$ or $z_T = nd^2/\lambda_{||}$, on optic axis, that is why the defocus series is necessary for recoding the correct high-resolution image. This is related to the 2D spatial periodic structural feature of the observed sample.

It may be interested that due to the de Broglie wave in electromagnetic wave fields, the same structure with different compositions should have different z_T because $z_T = (\lambda_d/n_r)/[1-(\lambda_d/n_r d)^2]^{0.5}$, where n_r is the refractive index of the medium. For electron microscopy, the defocus of the objective lens means the shift of the distance behind the sample as the sample is placed near the front focus plane of the ob-

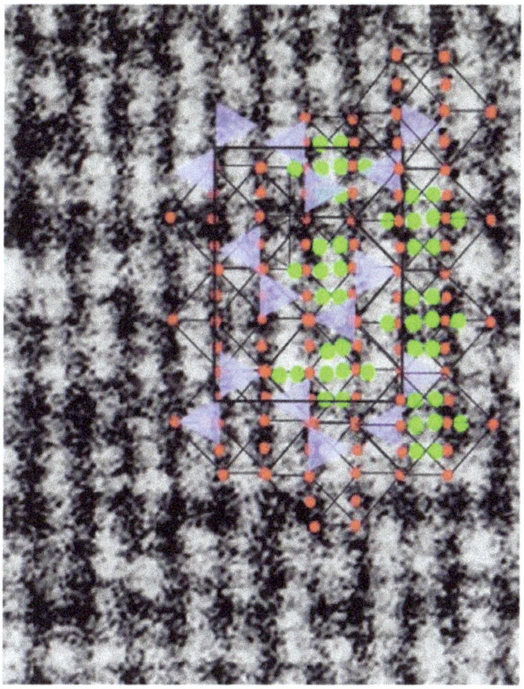

Figure 4.4: The TbO_x HREM image at [112] orientation. The white contrast shows the opening slits between (111) lattice planes and (112) planes, but the gap between (111) is wider than the gap of (112). The ordering oxygen vacancies have influence of the contrast between the (111) lattice fringes but not like Figure 4.3 clearly. This crystal is $Tb_{40}O_{72}$, and the model of its structure is inserted at the image.

jective lens. It is important to indicate that behind the sample it is the spatial domain in which the phase of the accumulated transient interaction, which is related to the particular track, holds and but only at the first Talbot distance, the larger spatial periodicity revived and the components of the 2D spatial periodicity of the structure would be appeared between the exiting surface of the sample and the first Talbot distance. Nowadays using series defocus of objective lens to get the Talbot self-images of Bragg tunnel grating of the sample is popular, but nobody seems to pay attention to understanding the pattern between the first Talbot plane and behind the sample.

As the flying time increased, the Talbot self-image smears out and the tilted orientation of the corresponding corpuscle electron pulses would be held and at the far field only the tilting partition of the ensemble flying corpuscle electron pulses would dominate the ensemble of the flying corpuscle electron train resulting in the diffraction pattern, as shown in Figure 4.2. For transmission electron microscopy at the back focal plane of objective lens, the electron diffraction pattern, which only carried the information of angular dispersion, existed. This back focal plane of the objective lens is a spatial frequency space. Due to the tilting of orientation of group velocity of a corpuscle electron characterized by spatio-spectral phase $\phi(x, y, \omega) = \eta x(\omega - \omega_0)$, the tilting magnitude is η. This phase varies linearly in frequency difference, and $\partial\phi/\partial\omega = \eta x$ corresponds to a delay in time domain and modified with the traveling tracks, where a slope varies linearly with frequency. If the spatio-spectral phase is $\phi(x, y, \omega) = \eta x^2(\omega - \omega_0)$, then the local wavefront will be curved. This indicates that different frequency components of the ensemble corpuscle electron pulse beam or train will evolve differently during its propagation. Therefore, the beam's spatio-spectral and spatio-temporal properties would change with its propagation. Based on the time interval, Δt, between the consecutive flying corpuscle electron pulses, the spatial-spectral phase, ϕ(x,y,ω), of the ensemble corpuscle electron pulses would be modified as its traveling in the slit or chink of the grating of a sample, while the individual corpuscle electron pulse is out of the grating, the accumulated spatial-spectral phase change would be fixed. In other words, the partitioned group velocity would hold and the beam front of the incoming ensemble of corpuscle electron pulses would disperse in different orientations, which correspond to the frequency difference, as the partitioned sub-ensemble group of the flying corpuscle electron pulses, which is called the electron diffraction beam. In the near field, which is behind the exited surface of a sample, different frequencies have different tilts, which result in varying best focal positions in the transverse dimension, usually called "transverse spatial chirp." This is the result of the transformation of the ensemble corpuscle electron pulses due to its propagation, which is the Talbot self-image formation. As the propagating time gains, the corpuscle electron pulse at focus no longer has any tilting but has longer local scales corresponding to the global duration in the near field. The focus point is spatially larger than the paralleling beam spot because different frequencies are focused at different transverse positions. Only at the near field, it is the spatio-spectral phase that differentiates them. Therefore, for detecting the atom's information in a sample by electron microscopy should be the near field of the transmitted ensemble of flying corpuscle electron pulses, which is the only place to be measured. The popular HREM images only exhibit the spatial periodicity of the Bragg grating, and explore the linear chirp induced by the interaction between the traveling corpuscle electron and atoms of the slit of grating. Using objective aperture as the mask of back focus of the objective lens can only obtain the spatial periodicity of the Bragg grating of the sample, but not the whole information of interaction between the traveling corpuscle electron pulse and the atom constructed the Bragg grating of the sample.

For objective lens, the diffraction spots on the back focal plane of the objective lens are the sources of rays to form the image at the image plane of the lens, and the aperture or mask of the objective lens determines the image feature. The diffraction spots, which are the assemblies of flying corpuscle electron pulses with different tilting angles of velocities for different spots and emanated from the Talbot self-image would form the image with corresponding spatial frequencies at the image plane. The aperture or mask at the back focal plane can select the diffraction spots to form different images corresponding to the spots. Because the diffraction in the Fraunhofer region (at the back focal plane) is linear and shift variant, it cannot be described by an impulse response or transfer function of the sample, but the Fraunhofer image can be modified by the transmitting function of the lens. The image formed by diffraction spots only exhibits the spatial frequencies of the sample or spatial periodicity in the sample. We emphasize that (1) as Bach has done while turning off the objective lens and using the intermediate lens to image the near field to get the intensity distribution map of the transmitted ensemble of corpuscle electron pulse beam; (2) while the objective lens is on, without objective aperture, the near-field intensity spatial distribution of the ensemble of corpuscle electron pulse beam is obtained; and (3) using the objective aperture as mask, the image obtained is the Fraunhofer image that dominated only by the objective aperture.

4.4 Converging corpuscle electron pulsed beam and narrow (nano) corpuscle electron pulsed beam

In the previous section, we discussed the incoming corpuscle electron pulse beam with the large transversal section L or $\kappa_L = 2\pi/L$ and stable consecutive flying temporal interval that results in the coherence of the beam would be same as the coherence of the individual corpuscle electron. The section of the transversal beam could cover many slits of the atoms grating and after transversing the atoms grating the incoming corpuscle electron pulses beam would be repartitioned into sub-beams with different propagation orientations and induce the periodic density distribution of the outcoming repartitioned corpuscle electron pulses beam as the Talbot image of the atoms grating at transversal section being perpendicular to the propagating direction in the near field of the sample. These repartitioned sub-beam densities are the observed objects of the objective lens in the transmission electron microscope. The flying corpuscle electrons in the density region have divergent group velocities and at different density regions there are different spatial and temporal frequencies. However, the objective lens and series imaging lens system of transmission electron microscope can only magnify the spatial frequencies but not the temporal frequencies, which means the corpuscle-carried temporal information uses other detection systems such as X ray energy-dispersive and electron energy loss spectra.

The observed images in the transmission electron microscope are not directly the electron density, but it is the luminous of displaying particles or molecules on a screen (or CCD). An individual corpuscle electron can be detected and the carried spatial and temporal information of the corpuscle electron may be collected. The transmission electron microscope uses the series magnetic lenses to amplify the space scale to $>\times 10^7$, if the pixel for detecting single electron can be smaller than 10^{-7} m, and the observed data should be the same as the images of HREM. This is what we expect in the twenty-first century.

4.4.1 Phase of corpuscle electron in convergent beam

As discussed previously, the single corpuscle electron is a confined dynamic electromagnetic energy flux in a corpuscle with space scale as $\lambda_{comp} = \hbar/mc$. The phase of an electron is

$$\begin{aligned}\phi &= 1/\hbar(\varepsilon_v t - px)\\ &= 1/\hbar\left[\varepsilon_0/(1-(v^2/c^2))^{-1/2}\right]\left[t-(v/c^2)vt\right]\\ &= (1/\hbar)\left[\varepsilon_0/(1-(v^2/c^2))^{-1/2}\right](1-v^2/c^2)t\\ &= (\varepsilon_0/\hbar)\left[(1-v^2/c^2)^{1/2}\right]t = (\varepsilon_0/\hbar)t' = \omega_0 t'\end{aligned}$$

The assembly of the flying corpuscle electron doughnut (or torus) pulses may be expressed as follows:

$$E(x,y,z) = \exp\left\{(\varepsilon_0/\hbar)\left[(1-v^2/c^2)^{1/2}\right]t\right\}E_0(z-v_g t) = \exp\{\omega_0 t/\gamma\}E_0(z-v_g t)$$

This means that in the laboratory frame and at any moment, a corpuscle electron doughnut pulse contains different frequencies, ω_0/γ, which is closely related to the velocity v or $\beta = v/c$. The time axis is the optical axis or z-axis in the microscope. The individual corpuscle electron doughnut pulse could contain different frequency distributions due to the velocity orientations of the corpuscle electron have different angles to the z-axis. Therefore, the angular divergences with the z-axis will dominate spatial and time frequencies of the individual corpuscle electron doughnut pulse. In other words, the bandwidth of the corpuscle electron doughnut pulses would be broad but the center frequency is still ω_0. Therefore, the convergent electron beam has broadband spectrum of spatial and temporal frequencies that may obtain more information about the interaction process between the sample and the traveling beam.

4.4.2 Robust single-electron scanning probe with single-electron detector would open a new field of quantum electron microscopy

Now it is available to control the electron emission from the nanotip emitter of an electron gun by the pulse of laser beam and numbers of electron in a bunch, which contain equal temporal interval between flying electrons, which imply that the spatio-temporal frequency bandwidth of the electron bunch may be same for the flying corpuscle electrons in the bunch. The transversal scale of the corpuscle electron bunch may be smaller than nanometer as a probe or elliptical cone with broader bandwidth. This probe may be scanning the sample with 2D having equal spatial and temporal intervals. This scanning mode is the Crewe mode of scanning transmission electron microscopy. If there is a single-electron detector that is available, then it is possible to collect the electron density distribution map of escaping electrons beyond the sample, which is the near-field image of the single-electron pixel digital detector. The shapes of incoming corpuscle electron doughnut pulse bunches may be manipulated by the magnetic lens or electromagnetic field such as laser beam and pulses that may provide the ability to measure or exhibit the dynamic state of the Poynting energy waves of electromagnetic field in the corpuscle electron frame after interaction with the sample. This is the ongoing revolution of transmission electron microscopy or the quantum electron microscopy.

As mentioned before, current transmission electron microscope is a huge instrument for making huge magnification by the series magnetic lenses, for example $>\times 10^7$. However, if we reduce the element scale of collecting impact electron energy to be 10^{-7} mm, the collected data should exhibit the same information as magnification images. Of course, that will be true based on two things: first is understood as what is an electron and what is the duality of particle and wave of an electron, and second is the developing single-electron detector with nanoscale pixel arrays.

We discussed the duality of a particle and wave of an electron in previous chapters and is summarized as follows: in Figure 4.5, we have fused different models and theories about an electron to give the electric and magnetic fields of a corpuscle electron doughnut pulse and electric intensity energy distribution $|E|^2$ in spatial periodicity (or wavelength) at corpuscle electron frame, then we proposal a dynamic spatial and temporal model at the corpuscle electron frame. Finally, we give a spatial contour of an electron at laboratory frame. From this figure we may understand that the corpuscle electron doughnut pulses at the corpuscle electron frame may be the confined vorticity field of electromagnetic energy flux waves in which the energy contains three parts: (a) the energy of point vortices in a classical inviscid, incompressible fluid contained in the elementary cell with a constant metric, may call “flat” torus; (b) the curvature energy represents changes in the kinetic energy due to the curvature of the torus induced by the locally distorted metric and would not be present for a flat metric; and (c) the quantum energy emanated from pure quantum contribution.

The electromagnetic energy flux may be expressed as the Poynting electromagnetic momentum flux as $S = \varepsilon/c = 1/c(E \times B)$. The periodic rotating motion of the electromagnetic momentum flux on a Möbius strip loop would have a confined electromagnetic field entity and the Poynting momentum wave would be as follows:

$$\mu_0[(1/c)\partial^2 <S>/\partial^2 t - \nabla^2 <S>] = -4\pi[\rho_m \ \nabla xE - \rho_e \ \nabla xB + 1/c^2(J_e xB + ExJ_m)] + \mu_0 \ \nabla x\nabla xS$$

If the Poynting momentum flux would be a wave, then the equation at the right-hand side would be zero. Therefore, the Poynting momentum wave field would be the vorticity field, as discussed in Chapter 1. The dynamic electromagnetic vorticity fields of a corpuscle electron result in the electric field and spin magnetic moment and inert mass of an electron in the laboratory frame, as shown in Figure 4.5. The quantum energy part of a corpuscle electron doughnut pulse would be the quantized vorticity as hω that can be the gain or loss during the interaction between the electron and external fields. The variation of the vorticity field of the Poynting wave depends on the spatial and temporal scales that can be understood by the phase of the vorticity field. Based on the de Broglie proposal $p = hk$, the $mv = hk \rightarrow v = (h/m)k = (h/m)\nabla\Phi$ if $\psi = E_0 \exp(i\Phi)$. Therefore, the velocity field of a flying corpuscle electron doughnut pulse would relate to the gradient of phase of the vorticity field. The Poynting momentum vorticity field will rectify the velocity direction of the electron during interaction between the scattering objects and flying electron. Measuring or detecting the electron displacement of a flying electron after traveling over the sample would get the sample's spatial-temporal information. The spatial and temporal metric scale of the interaction will dominate the final results that require scanning, and the recording rate should be matched. As discussed previously, controlling the emission of an electron emitter may obtain electron bunch with different number of electrons, which create the corpuscle electron doughnut pulse branches with variable bandwidths of spatial-temporal frequencies. After traversing the atomic group,

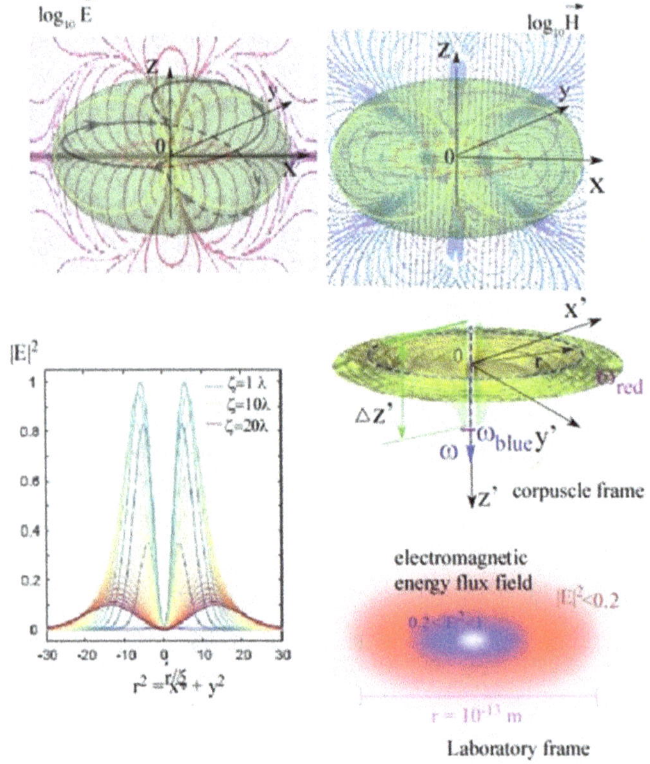

Figure 4.5: The electric and magnetic fields in a torus and the electric energy distribution in the torus. The model of vorticity Poynting field at the corpuscle electron frame and the electromagnetic field of an electron at the laboratory frame.

the outgoing corpuscle electron doughnut pulse bunch should carry the information of interaction with different corpuscle electron doughnut pulses in frequency and spatial domains shown in the single-electron detector device.

The single-electron detector devices have been developed by several groups, and in the near future, it is expected that the transmission electron microscopy would rapidly move from an era where there was limited information in the back focal plane to the one in which this information is recorded in detail on a fast pixelated single-electron detector that will provide additional and previously inaccessible information about the sample at a high spatial and time resolution.

As Professor A. Howie indicated, the transmission electron microscopy will continue skirmishing on the wave-particle frontier. The electron is a quantized confined electromagnetic vorticity Poynting field, which obeys the Maxwell and Newton laws in space and time. To analyze the physical information of spatial (or momentum) and temporal (or frequency) states in a corpuscle electron frame would be a great challenge.

References

Zdagkas, A., Papasimakis, N., Savinov, V. and Zheludev, N.I. "Building blocks for space-time non-separable pulses." arXiv:1912.09332v1 [physics. optics] (19 Dec 2019).

Zdagkas, A., Papasimakis, N., Savinov, V., Dennis, M.R. and I, N. "Zheludev singularities in the flying electromagnetic doughnuts." http://dx.doi.org/10.5258/SOTON/XXXXX.

Sanz, A.S. and Miret-Artés, S. "A causal look into the quantum Talbot effect." arXiv: quant-ph/0702224v2, (19 Jun 2007).

Hayrapetyan, A.G., Grigoryan, K.K., Götte, J.B. and Petrosyan, R.G.. "Kapitza–Dirac effect with traveling waves." New Journal of Physics. 17, 082002, (2015).

McMorran, B., Agrawal, A., Ercius, P., Grillo, V., Herzing, A., Harvey, T. and Linck, M. Origins and Demonstrations of Electrons with Orbital Angular Momentum, Philosophical Transactions of the Royal Society A-Mathematical Physical and Engineering Sciences, [online], (2017). https://doi.org/10.1098/rsta.2015.0434.

Azizi, B., Amini Sabegh, Z., Mahmoudi, M. and Rasouli, S. "Tunneling-induced Talbot effect." Scientific Reports. 11, 6827. (2021).

Chinosi, C., Della Croce, L. and Funaro, D. "Rotating electromagnetic waves in toroid-shaped regions." arXiv:1002.1206v1 [physics.class-ph] (5 Feb 2010).

Hestenes, D. "The Zitterbewegung interpretation of quantum mechanics." Foundation Physics. 20(10), 1213–1232, (1990).

Chaussarda, F., Rigneaultb, H. and Finota, C. "Two-wave interferences space-time duality: Young slits, Fresnel biprism and Billet bilens." Optics Communications. 397, 31–38, (2017).

Gilson, G. "Unified theory of wave-particle duality, the Schrödinger equations, and quantum diffraction." greyson.gilson@mulithinc.com

Zhong, H., Zhang, Y., Belić, M.R. and Yanpeng, Z. "Generating Lieb and super-honeycomb lattices by employing the fractional Talbot effect." arXiv:1902.06564v2 [physics.optics] (21 Feb 2019).

Walmsley, I.A. and Dorrer, C. "Characterization of ultrashort electromagnetic pulses." Advances in Optics and Photonics. 1, 308–437, (2009).

Bělín, J. and Tyc, T. "Talbot effect for gratings with diagonal symmetry." Journal of Optics, 20(2), 025604, (16 Jan 2018).

Bliokh, K.Y., Ivanov, I.P., Guzzinati, G., Clark, L., Van Boxem, R., Béché, A., Juchtmans, R., Alonso, M.A., Schattschneider, P., Nori, F. and Verbeeck, J. "Theory and applications of free-electron vortex states." arXiv:1703.06879v1 [quant-ph] (20 Mar 2017).

Deng, L., Hagley, E.W., Denschlag, J., Simsarian, J.E., Edwards, M., Clark, *.C.W., Helmerson, K., Rolston, S.L. and Phillips, W.D. "Temporal, matter-wave-dispersion Talbot effect." Physical Review Letters. 83, 26, (27 Dec 1999).

MacLaren, I., Macgregor, T.A., Allen, C.S. and Kirklland, A.I. "Detectors – The ongoing revolution in scanning transmission electron microscopy and why this important to material characterization." APL Materials. 8, 110901, (2020), doi: 10.1063/5.0026992.

Yessenov, M., Bhaduri, B., Kondakci, H.E. and Abouraddy, A.F. "Classification of propagation-invariant space-time wave packets in free space theory and experiments." Physical Review A. 99, 023856, (2019).

Kim, M.-S., Scharf, T., Menzel, C., Rockstuhl, C. and Peter Herzig, H. "Talbot images of wavelength-scale amplitude gratings." Optics Express. 20(5), 4903, (27 Feb 2012).

Mansuripur, M. "The Talbot effect." Optics & Photonics News. 8, 42–47, (April 1997).

Salem, M.A. and Caloz, C. "Space-Time Cross-Mapping and Application to Wave Scattering." arXiv:1504.02012v1 [physics. optics] (7 Apr 2015).

Janssen, M. and Mecklenburg, M. "Electromagnetic models of the electron and the transition from classical to relativistic mechanics." the proceedings of The Interaction between Mathematics, Physics and Philosophy from 1850 to 1940, a conference held at the Carlsberg Academy, Copenhagen, 26–28 Sep 2002.

Jhajj, N., Larkin, I., Rosenthal, E.W., Zahedpour, S., Wahlstrand, J.K. and Milchberg, H.M. "Spatiotemporal optical vortices." Physical Review X. 6, 031037, (2016).

Guenther, N.-E., Massignan, P. and Alexander, L.F. "Superfluid vortex dynamics on a torus and other toroidal surfaces of revolution." arXiv:1911.11794v1 [cond-mat.quant-gas] (26 Nov 2019).

Lindberg, R.R. and Shvyd'ko, Y.V. "Time dependence of Bragg forward scattering and self-seeding of hard x-ray free-electron lasers." Physical Review Special Topics-Accelerators and Beams. 15, 050706, (2012).

Deck-Léger, Z.-L. "Scattering in space-time abruptly modulated structures." Thesis. UNIVERSITÉ DE MONTRÉAL. (2017).

Chapter 5
Summary

As Professor A. Howie said, "In his (Ruska's) Nobel Prize acceptance speech, Ernst Ruska admitted that he became aware of the wave properties of the electron only as late as 1931 when he had already made his invention." This tells us that the success of transmission electron microscopy is based on electron particle optics, resulting in the advanced aberration-corrected transmission electron microscope. However, using the theory of "optic wave imaging" to interpret the recorded image is not coherent with the electron's physical nature. It is necessary to fuse the electron's physical nature with "wave imaging" to understand the imaging process, which requires knowledge of what an electron is. The earliest book on transmission electron microscopy was written by R.D. Heidenreich, titled *Fundamentals of Transmission Electron Microscopy* (Interscience, New York, 1964). In the book, he discussed "mass-thickness contrast." The second book is by P.B. Hirsch, A. Howie, D.W. Pashley, R.B. Nicholson, and M.J. Whelan, titled *Electron Microscopy of Thin Crystals* (Butterworths, London, 1965). In this book, they discussed "diffraction contrast." The third book is by John W. Cowley, titled *Diffraction Physics* (North Holland Publishing Co., Amsterdam, 1975). The fourth book is by J.C.H. Spence, titled *High-Resolution Electron Microscopy* (3rd ed., Oxford University Press, New York, 2003). The last book is by Ahmed H. Zewail and John M. Thomas, titled *4D Electron Microscopy: Imaging in Space and Time* (Imperial College Press, 2010). All these wonderful books expound on the observed phenomena but are not based on what an electron is and the particle-wave duality of the electron.

In the twenty-first century, what an electron is has been understood profoundly. Nowadays, an electron is understood as intrinsically spatial-temporal confined cycling electromagnetic vorticity fields of a photon with energy $\hbar\omega$ or mc^2, for which the vorticity field train of a photon is described as a helicoid in a twisted ribbon traveling electromagnetic wave geometric model. A closed-loop double-loop Hopf strip may be formed from a ribbon with a twist. An eccentric hula-hoop motion of a Hopf strip generates a swept volume toroidal envelope corresponding to a closed-loop standing wave. An electron consisting of two orthogonal spinors is generated by a rotating Hopf link corresponding to a poloidal and toroidal current loop. The charge trajectory is described by a rotating Hopf link, the simplest form of a knot, embedded in a torus manifold as discussed in Chapter 1. It is remarkable that an electron, which has a spin $1/2\hbar$ with quantized electric charge (e^-) and positive rest mass $\left(m_e = 0.511\,\mathrm{MeV}/c^2\right)$, may be created from an energetic photon that has a spin $\pm\hbar$ without electric charge and rest mass. The confined space of the helical path of a photon field is in the form of a Hopf link, which under rotation traces out the trajectory path of toroidal geome-

https://doi.org/10.1515/9783111449333-005

try. The electron has both the toroidal spin angular momentum $\hat{s}_0$ and the poloidal spin angular momentum $\hat{s}_r$, and the orientation of one vector would be independent of the other.

Based on this model, the electron is a fundamental elementary spatial-temporal cycling system of nature. The spatial profile is created by the dynamic rotation of the cycling photon on the double closed-loop Hopf strip, and the time flowing on the Hopf strip would continue infinitely. Based on the dynamic principle for each toroidal rotation, there are two internal rotations: toroidal and poloidal rotations. The toroidal radius corresponds to the reduced Compton radius, $R_c = \lambda_c/2\pi$. The tangential velocity of rotation v_t equals the velocity of light, c, which means $v_t = c = R_c\omega_c$. Because of the increased magnetic field, the orbital charge velocity internally varies from superluminal at the spin center to sublight velocity at the orbital periphery. The $1/2\hbar$ spin characteristic of the electron arises from the ratio of the Compton and Zitterbewegung rotational frequencies, $\omega_c/\omega_{zbw} = 1/2$, resulting in the observed net spin $1/2\hbar$ in a rest reference frame of the observer.

For transmission electron microscopy, the observer screen or CCD device is the rest frame in which the flying electrons form the dynamic electron current as an ensemble of corpuscle electron electromagnetic pulse trains with spatial-temporal order. Each corpuscle electron pulse is a "torus" pulse with nonseparability of space-time as $\xi^2 = p^2c^2 + (m_0c^2)^2$, where ξ is the total energy of a flying electron, p is its momentum, and m_0c^2 is the energy of an electron at rest state. The m_0c^2 may be expressed as $m_0c^2 = \hbar w_0$. Based on the de Broglie relation $p = \hbar k = \hbar/\lambda_{dB}$ and Compton wavelength $\lambda_{com} = \hbar/m_0c$, we can say that the toroidal dynamic cycling period is related to the de Broglie wavelength, but the poloidal vorticity motion frequency (the Zitterbewegung rotation) may relate to the Compton wavelength. It is clear that the de Broglie wavelength is larger than the Compton one as $\lambda_{dB} = (c/v_t)\lambda_{com}$. Therefore, we have the relation $\lambda_{dB} = (c/v_t)\lambda_{com}, \lambda_{com} = \times 10^{-13}$ m (for 200 kV $\lambda_{dB} = 1.45 \times 10^{-3}$ nm), which means the de Broglie wavelength dominated by the flying velocity, which is a spatial-temporal function as we discussed previously. For HREM, the high magnification (e.g., $\times 10^7$) of the transversal scale makes the energy detector pixel (e.g., a pixel = 10^{-5} m) of the CCD device able to collect single energetic corpuscles with super high frequencies (10^{20} Hz) consecutively. For a transmission electron microscope, we could steer the assembled profile of the flying corpuscle electron torus pulses by the magnetic lens series directly, but not the dynamic periodic motion of the internal circulation of the vorticity field of the photon. Therefore, the intensity distribution of the assembled profile of the flying corpuscle electron torus pulse train or beam would be what we observe at the microscope. This is one thing.

However, while the ensemble of flying corpuscle electron torus pulse train or beam transmits a sample such as an atom or grating in a microscope, the individual corpuscle electron torus pulse will interact with the sample. This interaction is dominated by the spatial electromagnetic fields and the propagation time duration. Due to the energy distribution in a corpuscle electron torus pulse relating to the spatial location in the torus, which has the highest energy or highest frequencies at the center of the torus and the lowest energy or lowest frequencies at the circumference, the result is the energy loss of a corpuscle electron torus or recoil to deflect the velocity orientation of the flying corpuscle electron torus pulse. This induces the intensity distribution of the assembly of the corpuscle electron torus pulses to be repartitioned, which is the spatial dispersion of the ensemble flying corpuscle electron called diffraction. This is another aspect of microscopy.

These two things together exhibit the wave-particle duality of the electron, which relates to the circulation coupled with linear motion, or space coupled with time, or field coupled with matter. The fundamental model of the electron tells us that the existence of discrete electrical particles manifests itself as a characteristic quantum phenomenon, namely as equivalent to the fact in which matter waves only appear in discrete quantized states.

That is what I understood: the success of transmission electron microscopy is due to the advantage of electron optics to create spatial resolution at the nanometer scale in the spatial domain and nanosecond scale in the time domain, along with the progress of Fourier imaging optics to explore the frequency and spatial domains. The quantum transmission electron microscopy, which is based on the dynamic circulation with linear motion of a confined electromagnetic vorticity field of a photon, may push forward the electron microscopy to decipher the encoded information of a single corpuscle electron in its internal frame after interaction with an atom.

It can be expected that in the twenty-first century, transmission electron microscopy will rapidly move from an era where there was limited information in the back focal plane to one in which this information is recorded in detail on a fast pixelated single-electron detector. This detector will provide additional and previously inaccessible quantum information about the sample at high spatial and time resolution. These quantum processing imagers will generate massive amounts of data due to increased speed, larger areas covered, ever-smaller pixels, and more information coming from each pixel. A similar trend toward ever-smaller, ever-more numerous pixels has improved the quality of imaging in the visible spectrum, allowing small details to be recognized even in large pictures. The deluge of "big data" resulting from the tiny pixel single-electron detector has led to the need for significant additional electronic activity in processing, storage, and transmission, instead of the huge magnification of magnetic lens system. The quantum electron microscopy would be a revolutionary change.

I greatly appreciate Professor John W. Cowley and Professor Leroy Eyring for giving me the chance to work and rethink this wonderful field. I wrote this booklet to commemorate the 103rd anniversary of Professor Leroy Eyring and the 100th anniversary of Professor John Maxwell Cowley.

Author operates super-high-voltage electron transmission microscope.

Index

https://doi.org/10.1515/9783111449333-006

www.ingramcontent.com/pod-product-compliance
Lightning Source LLC
Chambersburg PA
CBHW081536220326
41598CB00036B/6457